Contents

List of Tables

List of Boxes

List of Figures

Foreword

According to the World Health Organization, food safety issues are an important challenge to the public health sector. Many cases of foodborne illness go unreported and unrecognized, yet this type of illness is a significant contributor to the burden of disease in less developed countries. In particular, foodborne diarrhea remains one of the most common illnesses and cause of death for infants and small children. Improving food safety is therefore closely linked to improving nutrition and health outcomes, and must be part of any strategy to reduce poverty and hunger.

At the same time, growth in agricultural trade and the commercialization of biotechnology products brings increased international attention to food safety issues. Managing food safety risks is becoming a prerequisite for participation in international trade, and taking advantage of trade opportunities is an important element the World Bank's strategy to reduce poverty. There is an increasing realization that exports are a critical component in rural economic growth. Thus, food safety has a dual role in poverty alleviation, as it is important to public health and to market development.

But devising a strategy to address food safety is not simple. There are many sources of food hazards, public capacity to address these risks is limited, and the benefits are often difficult to measure. At the same time, several global trends lead to increased complexity in food systems, including increased trade in fresh and processed foods, growing urbanization and increased demand for foods of animal origin, and associated changes in the way that food is produced, processed, and distributed. Thus, defining priorities for public investments in food safety is challenging.

Recongnizing the critical importance of food safety, the Rural Development Family of the World Bank commissioned Laurian Unnevehr—a professor in the department of agricultural and consumer economics at the University of Illinois—and Nancy Hirschhorn—who was seconded to the World Bank from the U.S. Department of Agriculture—to provide an overview of the issues for the World Bank in designing a strategy for food safety investments. These include the role of food safety in public health and in export market development, the kinds of project activities that make sense at different stages of development, and how to evaluate such project investments. Dialogue with client countries and other international agencies about food safety is underway in a series of regional workshops. Through this process of dialogue, the World Bank seeks to define a strategy that recognizes the need for setting priorities, making sustainable investments, fostering partnerships with the private sector, and cooperation with international agencies. This paper serves as a basis for that discussion.

Robert L. Thompson
Director of Rural Development

v

Abstract

Food safety is receiving more attention worldwide with a rising incidence of foodborne disease, concern over new potential hazards, and growth in agricultural trade. Investments to improve food safety in Bank client countries can contribute to the Bank's mission in two ways: (1) by reducing the burden of disease in developing countries; and (2) by removing barriers to fresh food product exports, which are one potential source of income for the rural sector.

Knowledge and best practices in this field are rapidly evolving. The management of food safety risks and their regulation has changed dramatically within the last decade in many industrial countries. A progressive food safety regulatory system should include the ability to address food safety from farm to table, the use of comparative risk assessment to prioritize public action, an emphasis on prevention rather than inspection, an open decision making process involving stakeholders, and evaluation of public health outcomes. As food systems develop, client countries should build a food safety regulatory system that incorporates these principles.

International trade disputes over sanitary and phytosanitary (SPS) measures influence the ability of Bank client countries to compete in export markets. While controversies over genetically modified organisms (GMOs) are in the spotlight, the long-established SPS issues are equally important for developing countries. Bank client countries need to evaluate their interest in the SPS agreement under WTO; participate more fully in international agencies responsible for harmonization; develop the capability to assess equivalence for process standards, which are increasingly used for fresh food products; and resist the imposition of inappropriate standards.

There are a limited number of Bank projects that relate to food safety and they demonstrate the difficulties of addressing this complex issue. Export promotion projects provide the clearest lessons regarding the importance of private investments, public provision of information and training, and the need for targeted infrastructure investments. Projects supporting domestic food safety monitoring have not addressed the larger issues of how to build a comprehensive system for food safety improvement.

Growing attention to biosafety and SPS trade issues may lead client countries to make food safety investments. As the Bank becomes involved, it can draw on many other institutions for specific expertise in implementing projects and can encourage good practices for interventions and regulation. The Bank can ensure that food safety investments are undertaken within the appropriate policy framework so that food markets can develop most efficiently.

Acknowledgments

The authors are grateful to the members of the Food Safety Working Group for the comments and guidance they provided while preparing this paper, especially Cees DeHaan, Judith McGuire, Lynn Brown, Steve Jaffee, Alex McCalla, Tanya Roberts, Clare Narrod, Klaus Von Grebner, Michel Simeon, Maureen Cropper, Kseniya Lvovsky, Meghan Dunleavy, Laurent Msellati, Ousmane Sissoko, Jeff Hammer, Jock Anderson, Rajul Pandya-Lorch, Salim Hayabeb, and Oliver Ryan.

Special thanks also go to Daniel Nche, who provided research assistance and developed the project database used during this study, and Azita Amjadi, who provided the trade data.

Abbreviations and Acronyms

BSE	Bovine spongiform encephalopathy
COI	Cost of Illness
DALY	Disability-adjusted life year
FDA	Food and Drug Administration, U.S.
GAP	Good Agricultural Practices
GMO	Genetically Modified Organism
GMP	Good Manufacturing Practices
HACCP	Hazard Analysis Critical Control Point
IFPRI	International Food Policy Research Institute
IPM	Integrated Pest Management
PPQ	Plant protection quarantine
RDV	Rural Development Department of the World Bank
SPS	Sanitary and Phytosanitary
SSA	Sub-Saharan Africa
VOSL	Value of statistical life
WHO	World Health Organization
WTO	World Trade Organization
WTP	Willingness to pay

1. Executive Summary

Why Food Safety?

Food safety is an issue of growing importance due to several worldwide trends in food systems. The growing movement of people, live animals, and food products across borders, rapid urbanization in developing countries, changes in food handling, and the emergence of new pathogens or antibiotic resistance in pathogens all contribute to increasing food safety risks. These issues are recognized in international trade negotiations under the WTO and in the FAO's Committee on World Food Security.

While food safety may be an issue of growing importance, it is important to put it into perspective for Bank client countries. Inadequate food safety is a significant contributor to the burden of disease in developing countries, and should be addressed as the food system develops and along with related investments in public health. Basic sanitation and water services may be a prerequisite for addressing many food safety hazards. Food safety is one of many issues in developing fresh food product exports, which account for about half of all agricultural exports from developing countries. Such exports can play a role in rural poverty alleviation, but will not be the only source of increased income in rural areas. Thus, food safety is a significant issue for public health and for export markets in developing countries. Setting priorities for investments in food safety will require careful evaluation.

Evaluating Investments in Food Safety

In evaluating investments to improve food safety, risk assessment can provide the basis for understanding the sources of risk and their consequences. This can better inform efforts to meet export market standards or the development of domestic food safety regulations. Risk management, which is the policy process of making decisions about where to reduce risks, must rely on risk assessment for guidance. Risk management is a political process, and will reflect public perceptions about risk sources. A process of risk communication that includes stakeholders can help to create policies that will be more easily enforced.

Cost-benefit analysis of investments in food safety is straightforward for export markets, but more difficult for domestic public health. The value of additional exports gained through access to new markets can be compared to the costs of private and public investments to improve food safety. For domestic public health investments, cost-of-illness estimates have been used in several countries and some Bank studies. These place a value on lost productivity from illness, including disability and death, and provide one way of comparing the benefits from reducing foodborne illness to the costs of investments to prevent or control hazards. Such cost-benefit analysis can aid in setting priorities for public interventions to improve food safety.

Evaluation of public food safety investments must also consider whether they replace current measures to avoid risk undertaken by consumers or producers. If public action simply substitutes for private action, then neither risks nor total costs of risk reduction are reduced. This kind of evaluation is difficult, but crucial within developing food systems, where the incidence of food safety risks is undergoing change.

Managing Food Safety and Quality

The costs and nature of food safety investments will be influenced by the growing recognition that a farm to table approach is necessary to address food safety. Because many hazards can enter the food chain at different points and it is costly to test for their presence, a preventative approach that controls processes is the preferred method for improving safety. Sometimes characterized as a Hazard Analysis Critical Control Point (HACCP) system, this approach is increasingly used as the basis for food safety regulation and for private certification of food safety.

Food quality issues are closely related to food safety in practice, because both are managed throughout the production process. Market failures can also occur for quality, and public intervention can sometimes address these failures through establishment of grades and standards. Because product quality may differ among markets and quality differentiation can benefit industry, private actions can often substitute for public actions to address issues of product quality. The public role in addressing food safety is much more clearly justified, because it protects consumers and promotes public health.

Lessons for Developing Food Safety Regulation in Bank Client Countries

Current understanding of food safety management and the desire of most industrial countries to be responsive to consumers and efficient in the use of public resources has brought about changes in food safety regulatory systems. A progressive food safety regulatory system includes:

- Consolidated authority with ability to address the food system from farm to table and to move resources towards the most important sources of risk.

- Use of comparative risk assessment as one criterion for prioritizing public action.

- Cooperation with industry and consumers to provide information and education.

- Use of HACCP principles to promote prevention and industry responsibility in place of prescription and inspection.

- An open decision making process that allows stakeholder participation.

- Evaluation of public health outcomes from regulation.

These elements provide guidelines for developing food safety regulation in developing countries.

Food safety interventions build from basic investments and simple interventions to more complex regulatory systems as economies develop. Priorities for public action change at different levels of development. At low levels of income, investments in water and sanitation and targeted interventions to reduce child malnutrition are the highest priority for food safety. As food systems and the capacity for food safety policy develop, then targeted interventions for single source hazards or important control points in the food marketing chain become practical. At higher levels of income, more extensive regulation and enforcement are feasible.

One of the difficulties that developing-country governments face is how to phase in interventions as the formal food sector grows, without driving out informal activities that still serve an important economic function. This tension suggests that an initial emphasis on risk prioritization, training, and provision of information is the right approach, rather than imposing standards and inspection. Information-based interventions help the entire food system without penalizing the informal sector, and accord with the principles emerging from food safety regulatory reform in industrial economies. Some economies of scale may be possible through a regional approach to these issues. The efforts of PAHO in Latin America and the Caribbean (LAC) and of ASEAN in Southeast Asia may provide some useful models and lessons.

An important issue is how efforts to improve export quality may provide benefits for domestic consumers. The foreign exchange generated from export markets can provide incentives for food safety improvements, e.g. through investments in processing facilities, that have positive spillovers for domestic markets. On the other hand, export standards may not be appropriate for domestic production, if domestic risks differ due to local handling and preparation. Understanding the priorities for domestic public health can inform policies that capture benefits for domestic consumers without creating barriers for export markets.

Meeting Sanitary and Phytosanitary Standards in Export Markets

Food export markets present a somewhat different set of challenges from domestic food safety regulation. Exports of fresh food products such as meat, fish, fruit, and vegetables represent a growth opportunity because these products have a high income elasticity of demand and fewer trade barriers than traditional agricultural exports. Fresh food products are more likely to encounter sanitary and phytosanitary barriers to trade. Delivering safe food to distant markets requires process controls throughout the production process and mechanisms to certify to buyers that such controls are effective. Developing-country exporters need to know how to meet standards in different markets and how to meet the increasing demand for product trace-back and certification of production methods.

The SPS agreement of 1994 provides a framework for resolving disputes about SPS measures under the WTO. There is evidence that this agreement has stimulated activity to reduce SPS barriers to trade, but there remains significant disagreement at the international level over the role of science and consumer choice in regulating risk. Controversies at the global level influence the ability of Bank client countries to compete in export markets. They create uncertainties about the potential acceptability of production methods and products in different potential markets.

Issues for Developing Countries in WTO 2000

How to regulate trade of potentially risky food and agricultural products will be a contentious issue in WTO 2000. The current disagreements regarding risk management at the international level could lead to a re-examination of the SPS agreement. This could bring about new trade barriers that might hinder growth and development for developing countries. Resolving these disagreements in a way that will not preclude trade expansion is a serious challenge for all of the participants in WTO 2000. Developing countries will need to evaluate their own interests in terms of trade opportunities and domestic risk aversion in order to participate fully in this debate. While GMOs and biotechnology will be in the spotlight, the long-established sanitary and phytosanitary issues are also important for developing countries.

If the framework of the existing SPS agreement remains intact, then there are three specific areas for developing countries to consider in preparing to participate. The first is the role of the Codex, the OIE, and the IPPC in setting internationally recognized standards. Many developing countries do not have the capacity to participate in these international organizations and it is not clear that their interests are always taken into account. Learning how to participate effectively involves capacity building within the public sector in developing countries and may also involve forming coalitions around issues of mutual regional interest.

A second issue is the growing use of process standards for food safety and the resulting difficulties in determining equivalence across countries. Equivalency in risk outcomes is the de jure standard under the SPS agreement. However, in practice, equivalency might be determined through requiring particular processes from particular countries. Whether risks really differ for commodities produced in developing countries must be examined carefully on a case by case basis. Developing countries need to participate in the current discussions regarding risk assessment and determination of equivalency that are proceeding within the Codex to ensure that they are not held to a de facto higher standard.

The third issue is whether a domestic food safety regulatory system is becoming a necessary prerequisite for participation in trade. Some components of a food safety system may be necessary, but developing countries should resist the trend in importing countries towards requiring equivalent systems rather than equivalent outcomes for specific products. Some policy analysis and risk assessment capacity is necessary for participating in international trade negotiations. Some control over animal and plant hazards, or monitoring of imports to prevent hazards, may also be necessary to enable trade participation. It is possible that such capacity can be developed at a regional level, if there are common trade interests and similar risks. A full scale legal framework for food safety regulation with minimum standards is expensive for small low income countries, and may be counterproductive in any country without sufficient public capacity for enforcement.

Lessons from Bank Project Experience

There are a limited number of Bank projects that relate to areas of food safety. The most closely related Bank activities have been in animal and plant health. The successful animal and plant health projects provide some models for functions that could be supported to improve food safety, including laboratory tests to monitor the food supply, mobilizing industry support for public efforts, and cost recovery for public activities that benefit the agricultural industry. But the goals of animal and plant health projects, and the constituencies, are fundamentally different from food safety. Animal and plant health increase productive capacity, and industry will therefore support public actions that cannot be undertaken by individual producers. Food safety is a public health goal that benefits consumers and may impose costs on producers. Therefore agencies with experience in animal and plant health issues may have some of the right technical expertise for food safety, but will not necessarily be oriented towards consumer interests.

The export promotion projects provide the clearest lessons. These include the importance of partnerships between the public sector and the private sector, the importance of training and market information, the close connection between safety and quality management in practice, and the need for small, but crucial, infrastructure investments or policy reforms. These projects and other studies of successful non-traditional food exports suggest that the best practice is to rely on private certification and private investments as much as possible. The strong market incentives in exporting should motivate both.

The domestic food safety regulation projects raise some difficult issues. The first is how such regulation will interact with other policy reforms. There is a tension between fostering a weak and potentially corrupt public role versus strengthening public capacity for an appropriate public activity. Resolving this tension will only occur on a case by case basis as opportunities to strengthen institutions and change regulatory approaches are identified.

The second difficult issue is how to relate project components to the overall elements of a progressive food safety regulatory system. Investments in laboratories alone do not bring about a comprehensive and effective system. Such activities should be linked to enforcement and to outcomes in food safety improvement. But more importantly, such investments need to be part of a larger effort that includes the legal framework for regulation, capacity building for comparative risk assessment and for HACCP approaches, and institution building for greater stakeholder involvement. Ideally, a comprehensive approach would also identify key infrastructure investments to support food safety through better sanitation at key points in the food chain.

Recommendations for Bank Action

Investments in food safety could contribute to the Bank's mission to reduce poverty and improve living standards. But such investments must be evaluated within the larger context of social and economic returns. In particular, larger issues of food security and food market reforms take precedence over efforts to improve food safety. Food safety regulation is likely to fail to achieve public health goals when markets are not performing well or when food price and trade policy distorts incentives. Similarly, improving food safety and quality in export markets cannot overcome barriers to exports from policies or inadequate infrastructure. Nevertheless, food safety regulation is a legitimate public activity because markets for food safety are often incomplete. It is the kind of public activity that can enhance market performance when undertaken in combination with price and trade reforms.

Growth in agricultural trade and the commercialization of biotechnology products are bringing increased attention to food safety in Bank client countries. When they occur, food safety project investments will be part of investments to support public health, agricultural productivity, or export promotion. The demand for project components is likely to differ across regions.

The Bank has limited experience in food safety issues, but has the leadership capacity to foster dialogue about SPS issues at the international level and regional partnerships to address issues of mutual interest. When the Bank undertakes food safety investments it can help to implement the good practices identified above. In designing and implementing projects, the Bank can draw upon many other institutions for expertise in food safety, including the FAO and WHO, bilateral assistance from the major industrial-country importers, and private sector organizations. The Bank's comparative advantage among these organizations is to ensure that food safety investments are undertaken within an appropriate policy framework so that food markets can develop most efficiently.

The Bank is in the process of defining a strategy for food safety investments through dialogue with client countries in a series of regional workshops. The first regional workshop, organized in partnership with EMBRAPA and IICA, and was convened in Costa Rica in August 1999 (IICA, 1999). Through this process of dialogue, the Bank hopes to define a strategy that recognizes the need for setting priorities, making sustainable investments, fostering partnerships with the private sector, and cooperation with other international agencies. This paper serves as a basis for that discussion.

2. Introduction

Food safety issues are receiving growing attention worldwide. Science is documenting new hazards and providing a better understanding of the scope of foodborne illness. Microbes are mutating and surviving in new food production and processing niches (Tauxe). Growing urbanization and increased global trade in food products pose new challenges to maintaining or improving food safety (Kaferstein and Abdussalam). Thus, questions about food safety issues will arise more frequently in the future. The Bank needs to consider how and whether to become involved in providing leadership for improving food safety in client countries.

This paper has two purposes. First, it provides background to better inform policy dialogue with client countries on this issue. Second, it explores how the World Bank and its client countries might invest in improving food safety. Such investments could potentially contribute to the Bank's mission to alleviate poverty, by improving human health and nutrition or by expanding earnings from food export markets. We review Bank project experiences and the emerging literature on the costs and benefits of food safety improvement to look for "best practices" or guidelines for investments.

Because Bank experience in this area is quite limited, most of the paper draws upon the literature to inform the strategic choices facing the Bank and client countries in addressing this issue. The issues include:

- What is the significance of food safety and how does it relate to the Bank's mission for poverty alleviation?

- How do food safety issues differ for export promotion and for domestic public health?

- How do debates about food safety risk management in international trade influence developing countries?

- What are the appropriate public and private roles in addressing food safety?

- What are the special concerns of the small, low income food insecure countries?

- What are the lessons from international experience, including Bank projects, regarding successful approaches to improving food safety?

- What are the implications for Bank activities?

What is Food Safety?

Unsafe food contains hazards that can make people sick, either immediately or by increasing their risk of chronic disease. Hazards that receive attention from policy makers include:

- Microbial pathogens—Microorganisms like *Salmonella, Listeria, Campylobacter*, or *E.coli* that occur naturally in animals, humans, or the environment can cause diarrhea and long term complications like kidney failure.

- Zoonotic diseases—Diseases like tuberculosis or brucellosis can be transmitted from animals to humans through food.

- Parasites—Intestinal worms afflict several hundred million people in developing countries; some are transmitted through contaminated water or food.

- Adulterants—Physical contaminants in food, such as metal or glass, or other non-food elements, such as rodent feces, can pose a hazard.

- Mycotoxins—Naturally occurring on plants or in animal products when animals eat feeds containing mycotoxins. Mycotoxins increase the risk of cancer in humans.

- Antibiotic drug residues—When an animal receives antibiotic drugs through feeds or improper treatment, residues can occur in animal products. Some residues pose chronic risks; antibiotic use in animals is also alleged to contribute to the growing resistance of microorganisms to antibiotics.

- Pesticide residues—Result from pesticide use in production and distribution. Some residues increase the risk of cancer; others carry risks of neurochemical damage.

- Heavy metals—Enter food through the soil or water and can cause acute or chronic illness.

- GMOs—Genetically modified foods may contain allergens or toxins that are not found in conventional foods.

While there are many potential foodborne hazards, the reported incidence of foodborne illness from microbial pathogens has been increasing worldwide. The World Health Organization reports a rising incidence of foodborne illness in industrialized countries (WHO). The potential causes of increased foodborne illness include:

- growing numbers of immune-compromised or elderly in the population worldwide;

- the emergence of new pathogens or of anti-biotic resistance in pathogens;

- changes in food handling, storage and preparation practices; and

- the growing movement of people, live animals, and food products across borders, which spreads disease more widely when it occurs.

It is now recognized that some foodborne diseases can lead to serious chronic health conditions, such as Guillan-Barre syndrome or rheumatoid arthritis (CAST). As the causes and long term health consequences of foodborne illness from microbial pathogens are better understood, industrial countries have strengthened their efforts to control these pathogens.

How Does Food Safety Relate to Animal and Plant Health?

Food safety differs from animal and plant health in its focus on foodborne hazards to human health. Usually animal and plant health improvement is tied to enhanced agricultural productivity. However, there are strong linkages among the three issues. First, some animal diseases can also be transmitted to humans, so improving animal health is sometimes linked to food safety. Similarly, some plant diseases or control measures involving pesticides have implications for human health through residues on food. Second, similar regulatory approaches may be used to reduce risks in all three areas, so designing public systems to address all three has some economies of scope. Third, improvements in all three areas may be a prerequisite for entering international trade, and thus may need to occur simultaneously. Fourth, all three areas fall under the SPS agreement and thus are addressed in the same way in international trade disputes. These linkages mean that some Bank experiences in animal and plant health may provide lessons for investments in food safety.

How Important are Food Safety Hazards in Developing Countries?

All countries share similar concerns about food safety hazards, but the relative importance of risks differs with climate, food habits, levels of income, and public infrastructure. Some risks are greater in developing countries. Poor sanitation and inadequate drinking water supplies pose a much greater hazard to health in developing countries than in developed ones. Diarrheal disease accounts for most illness and death in children under five (1993 World Development Report). According a WHO study,

> *Of the approximately 1,500 million global episodes of diarrhea occurring annually, resulting in 3 million deaths among children under five, 70 percent have been estimated to have been caused by biologically contaminated food. Contaminated food has been recognized as playing a major role in the epidemiology of cholera and other forms of epidemic diarrhea, substantially contributing to malnutrition (Motarjemi et al 1993).*

Other hazards may also differ by climate and level of income. Mycotoxins are more prevalent in the sub-tropics and tropics, and pose greater risks where diets are concentrated in foods that have higher mycotoxin levels (Bhat and Vasanthi). Parasites are also more common. For example, cysticercosis is endemic in rural areas of Latin

America, Asia, and Africa, infecting between 2 and 15 percent of the population, in comparison with less than one-hundredth of a percent infected population in the U.S. (Roberts et al).

Many of the reasons why food safety is becoming a more important issue worldwide are most compelling in developing countries. Increased foodborne disease can be expected from increased urbanization, which lengthens the food chain and introduces new risks; climate change, which encourages the growth of some pathogens; shrinking fresh water supplies, which increase the cost of risk mitigation; and the growth of immune compromised populations who are susceptible to foodborne illness. In addition, expanded food product trade and migration will contribute to the globalization of foodborne diseases. All of these changes are likely to be more dramatic in developing countries.

Why Should Bank Client Countries Invest in Food Safety?

Investing in improving food safety has two potential benefits for Bank client countries. First, such investments can facilitate growth in food product exports. Sanitary barriers to trade are becoming more important as traditional trade barriers are reduced and as world trade in food products grows. To participate effectively in international markets, developing countries need to be able to ensure the safety of exported and imported food products. Second, investments in food safety can improve the health and well-being of consumers in Bank client countries. Reducing the burden of disease can increase productivity and the quality of life. The discussion below is framed around these two outcomes: 1) export markets and 2) public health.

Investments in food safety could contribute to the Bank's mission to reduce poverty and improve living standards through sustainable, broad-based growth and investment in people. The specific contributions come under two of the four critical goals defined in the Rural Sector's *Vision to Action*. These are: 1) poverty reduction; and 2) household, national, and global food security. First, expanded food exports and trade can play a role in poverty reduction within the rural sector, through broadening income earning opportunities and allowing greater specialization in production. As discussed below, food safety investments may be a necessary condition for expanded food exports. Second, improvements in food safety within client countries can improve health and nutrition. Reductions in foodborne disease allow individuals to make better use of the calories they consume, thus effectively reducing nutritional requirements. Furthermore, this contributes to poverty alleviation through increasing the quality and length of life. Third, improvements in food safety foster food market development by increasing the scope for both domestic and international trade. This contributes to both national and international food security. However, investments in food safety must be evaluated within the larger context of alternative investments in poverty alleviation. Therefore, the paper begins with a discussion of how to evaluate the benefits and costs of food safety investments.

3. Evaluating Public Investments in Improving Food Safety

This section provides an overview of basic concepts that are now widely accepted in evaluating public investments and interventions in food safety. We begin by reviewing the rationale for public intervention and the policy analysis tools used to evaluate interventions. Then we turn to the areas where there is often disagreement and explain the sources of controversy in making food safety policy.

Market Failure

Many food safety hazards are not easily detectable. Neither food producers nor consumers are fully informed about many aspects of product safety. Food producers cannot guarantee a particular level of safety, and consumers cannot always effectively demand safety. Thus improved food safety is not always rewarded in the marketplace. Incomplete information about food safety is the justification for public sector intervention.

The degree of market failure varies widely in practice. In export markets, countries and industries have strong incentives to avoid food safety incidents, which can destroy product reputation. In domestic markets, consumers can exercise control over some hazards through food handling and preparation (e.g. thorough cooking). But this is not possible when they rely on food services or vendors for final preparation. In some markets, consumers use experience as a guide to identifying relatively safe products and suppliers. For some foodborne hazards, the effects may be chronic rather than acute or the exact source may be difficult to trace, given that symptoms often occur 3 to 5 days after food consumption. Thus experience can provide only limited guidance and markets will not always punish suppliers of bad food.

The existence of market failure is a necessary but not sufficient condition for intervention. Not all risks can be successfully addressed by intervention and it may not be economic to do so. Next we consider issues in risk management.

Risk Analysis, Risk Assessment, and Risk Management

Food safety involves hazards that occur with unknown frequency and may have unknown outcomes. Because there are uncertainties associated with efforts to improve food safety and because outcomes are difficult to measure, it is useful to apply a *risk analysis* framework in policy making.[1]

The first step is *risk assessment*, which includes hazard identification, hazard characterization, exposure characterization, and finally, risk characterization. The end result is a quantitative estimation of the adverse effects that are likely to occur in a given population, to be used in decision making.

This characterization provides the basis for *risk management*, which involves making decisions about where to reduce risks. *Comparative risk assessment* identifies the most important risks (i.e. those that affect the largest number of people or which have the most severe outcomes). Identification of where and how these risks are likely to occur will show what kind of interventions might reduce risk most effectively. Without this kind of information, efforts to reduce risk might waste resources by focusing on ineffective interventions or on hazards with low risks.

Risk management involves making decisions about acceptable levels of risk. Simple risk assessment does not provide answers to this question. Quantitative estimates of the value of risk reduction are possible, through estimating the value of human health and life. Such estimates can provide some guidance on risk priorities. But the level of acceptable risk will also depend on social norms and public perceptions. Thus risk management involves the art of combining quantitative assessments with social and economic concerns to make decisions about risk reduction.

The final step in this process is *risk communication*, which involves public education regarding what is known about hazards and their risks, the uncertainties associated with current knowledge, and the reasons for any interventions to reduce risk.

The process of *risk analysis*, founded on quantitative risk assessment, is widely recognized in industrial countries, and by the WTO as a basis for determining equivalency of sanitary regulations across countries. But it is not always put into practice in regulating food safety. It has been most widely applied to risks from pesticide residues on food. However, not all hazards lend themselves to quantitative risk assessment. In particular, risk assessment for microbial pathogens in food is in its infancy (Hathaway). Furthermore, less is known about some potential hazards (e.g. GMOs) so it is difficult to quantify their risks.

Given these qualifications, and the lack of risk assessment capacity for many of the hazards that are important in developing countries, how useful is the concept of risk assessment to Bank client countries? Developing capacity for risk analysis in developing countries is important for several reasons. First, consider the issue of meeting

---

[1] See American Chemical Society. *Understanding Risk Analysis*. 1998. Available on the web http://www.rff.org/misc_docs/risk_book.htm for a thorough analysis.

standards for food safety in food exports. The export market determines the level of acceptable risk, which is a given condition to exporters. But within any exporting country, a risk assessment can identify the most cost-effective way to reduce risk and can provide importing countries with assurances that products meet their standards. Because risks are likely to differ for developing country suppliers, they must be able to clearly identify hazards, risks, and cost-effective ways to reduce risks and meet export standards.

Next, consider issues of domestic food safety within developing countries. Adopting risk standards or priorities from industrial countries might not make sense, when hazards and risks differ. A good example is the balance between the risk of infectious agent for cholera and the potential risk of chronic effects of carcinogens from the water chlorinating process, which reduces the cholera causing agent. In the early 1990s, a decision by the Peruvian government not to chlorinate drinking water, due to potential cancer risks, led to a cholera outbreak (Anderson). Risk assessment, even of a qualitative nature, provides some guidance on the most important risks and where it makes sense to intervene. The Bank's commitment to address intestinal parasites reflects this kind of qualitative risk assessment. Interventions to eliminate this health hazard are inexpensive relative to the perceived benefits (Binswanger and Landell-Mills).

An important element in risk management is cost-benefit analysis of interventions to reduce risks. Risk assessment provides guidance on the potential value of interventions and some initial prioritization among risks and interventions. Next, we discuss how to evaluate a specific investment in food safety.

Evaluating Net Benefits from Food Safety Investments

Export Promotion

The net benefits of food safety investments for export promotion are the value of additional exports gained through access to new markets or the value of avoiding loss of reputation in existing markets. These can be compared to the costs of private and public investments to improve food safety and to certify that a standard is met. In chapter six, we discuss several Bank projects in export promotion that have generated returns of this kind.

An important issue for the Bank is the extent to which agricultural exports contribute to poverty alleviation. This issue was addressed in both the *The World Bank Strategy for Reducing Poverty and Hunger* and in the Rural Development strategy, *Vision to Action*. Both documents assume that growth in agricultural export markets will be one of many potential ways to reduce rural poverty. These strategies draw upon a series of multi-country studies carried out by the International Food Policy Research Institute (IFPRI) in the 1980s regarding the impact of commercialization of agriculture on the rural poor (summarized in Binswanger and Von Braun). Commercialization generally benefits the poor through increasing incomes and job security. In order for the benefits from commercialization to be widely distributed, it must be accompanied by policies that provide equal access to markets and resources. A detailed discussion of such policies is beyond the scope of this paper.

Domestic Health and Welfare

The net benefits of food safety investments in the domestic market are much harder to estimate. It is often impossible to observe a market value for food safety, because investments in food safety may or may not be rewarded in the marketplace with higher prices or expanded demand. Such investments reduce the incidence of disease and death. In industrial countries, economists have used a variety of non-market valuation techniques to estimate the social welfare improvement from reducing food safety hazards (Caswell). Cost-of-illness (COI) estimates are easy to communicate to policy makers and have been widely used to justify food safety regulation in many industrial countries (see box 1). These estimates measure medical costs and the value of lost productivity from illness and premature death.

Although COI estimates have many advantages, economists have raised concerns about their use in cost-benefit analysis of public interventions. First, COI estimates may represent only a part of the true cost of foodborne illness to consumers. Because they measure the cost of actual illness outcomes, they do not measure the full value of

---

Box 1   Cost of illness estimates for valuing investments to improve health

Cost of illness estimates have been used to value food safety in the U.S., Canada, Great Britain, and Australia. This methodology considers two components of the costs of foodborne illness: lifetime medical costs and lost productivity (e.g. forgone earnings). If foodborne illness were reduced, these resources could be reallocated for other uses.

Medical costs include physician and hospital services, supplies, medications, and special procedures unique to treating each foodborne illness. Medical costs reflect the number of days/treatments of a medical service, the average cost per service, and the number of patients receiving such service.

Productivity losses occur when there is a reduction or cessation of work due to premature mortality and morbidity. Although most people with foodborne illness miss only a day or so of work, some die or contract such physical complications that they never return to work or regain only a portion of their pre-illness productivity. Average wages or salaries provide one measure of lost productivity.

The productivity losses for illness-related death are estimated in several different ways that usually encompass more than just foregone earnings. The value of a statistical life (VOSL) may be estimated from studies of willingness to pay to reduce risks, the value of human capital, or some combination. Viscusi summarized the results of several studies of how wages increase with job risks, and estimated that the VOSL in the U.S. ranges between $3 million and $7 million in 1990 dollars. In the U.S., FDA used a VOSL of $5 million in evaluating the benefits of mandating HACCP for the seafood industry.

Source: Buzby, etal.

---

risk reduction. It has been argued that consumers value a reduction in the risk of illness, even if they would not have become ill (Cropper).

Another concern is about valuing consumer's current efforts to mitigate risk (Van Ravenswaay and Hoehn). If public interventions replace consumer actions, is there a net benefit for society? So, for example, if mandatory pasteurization replaces boiling milk in the household, does this just shift the costs of risk reduction? Or does it result in reduced risks and reduced costs of risk reduction overall? A full cost-benefit analysis would measure only the net reduction in risk and the net change in costs from any intervention to improve food safety.

These controversies are relevant to Bank client countries because food safety is often assumed to be something that only the rich can afford. In other words, it is assumed that consumers must bear the costs of unsafe food (either risk mitigation or illness outcomes), and that safer food is a luxury good. This argument stems from observations of emerging demand for process characteristics (e.g. organics, animal welfare) in high income countries. However, as discussed above, food safety hazards of many kinds are important in developing countries. Furthermore, these hazards have important consequences for the poor. The costs of lost productivity may be devastating for the poor, who depend on their labor for income and may not have access to any formal medical care to mitigate possible serious outcomes. The costs of mitigation measures to reduce risk (e.g. boiling water) may be beyond the reach of the poor, which might justify public intervention on their behalf. The issue is then how to measure the benefits of improved food safety in a poor country.

The Bank has used both cost-effectiveness (cost per life saved) and cost of illness approaches to value health projects (Hammer). Examples are found in a recent Bank study of the benefits from sanitation in Andra Pradesh (Lvovsky et al). The study measured the burden of disease in terms of the value of a disability adjusted life year (DALY), which measures both loss of life due to premature death and fractions of healthy life lost as a result of illness or disability. The value of a DALY was based on willingness to pay (WTP) to reduce risks, which was assumed to be three to six times GDP per capita. This is very similar to the value of life used in cost of foodborne illness estimates, which rely on hedonic wage studies of workers' willingness to accept risky jobs in developed economies. Adjusting the willingness to pay to avoid risk according to GDP per capita allows comparable risk valuation for allocating public funds within an economy at a particular stage of development. The Andhra Pradesh study also examined marginal benefits and costs from sanitation investments and the cost per DALY saved from different kinds of investments. These comparisons of both benefits and cost-effectiveness provided further guidance to choose among investments that promote health.

Thus, the cost of illness approach is likely to be the most straightforward way to value benefits from food safety investments. It focuses evaluation on illness outcomes and on how widespread or serious these outcomes are likely to be. The difficulty of applying this methodology is that it requires data on the incidence of illness and illness outcomes such as death and disability. Because such data are scarce and unreliable, a qualitative ranking of investments may be the best evaluation possible or specific studies of disease incidence may be necessary to evaluate interventions. As discussed below, the effects of replacing consumers' risk mitigation efforts should be taken into account so that only net benefits and changes in costs are measured (Hammer).

Costs of food safety interventions are generally easier to estimate than benefits, since they include the observable costs of actions by producers and consumers to reduce risk. Such costs can include investments in new equipment or the labor and other inputs associated with increased sanitation or care in handling. Public sector costs include those associated with monitoring and enforcement.

Certain kinds of food safety interventions require large fixed capital investments, including both equipment and human capital, which are easier to justify for larger firms (MacDonald and Crutchfield). Therefore, mandating some kinds of food safety controls may impose greater costs on small firms, and encourage industry concentration. This concern is especially relevant to developing food systems where many small firms may co-exist with larger ones, and we discuss the implications for phasing in regulation in developing countries in chapter four.

Public Interventions to Improve Food Safety

Table 1 shows the types of possible interventions to improve food safety. Command and control (CAC) type interventions set standards and requirements for food product suppliers. Incentive-based interventions provide better market information so consumers can more effectively demand greater safety.

Table 1 Possible interventions to correct market failures in food safety

Type of Intervention	*Examples*
Command and control type	
Process Standards	Specifying how products are produced, e.g. GMPs or GAPs; including pre-inspection in countries of origin for imports
Outcome Standards	Testing and inspection to ensure that products meet a particular safety standard, incl. monitoring food imports
Incentive-based type	
Providing information to the public	Foodborne disease surveillance; informing consumers about how to avoid risk; training industry in HACCP procedures.
Facilitating Private Bargaining	Labeling products that may either contain hazards or have been produced under certified control measures.

Source: Unnevehr and Jensen (1996).

Incentive based interventions allow consumers and producers to respond to market signals. The market will determine the optimum level of safety. However, incentive-based interventions may not be possible if information is just as costly for public agencies as for private actors. Or, society may wish to protect certain vulnerable consumers (e.g., children), and thus prefer a command and control type intervention. Outcome standards are usually preferred to process standards, because they allow firms to meet the desired standard with the least cost production method. Process standards are more prescriptive in terms of telling food suppliers how to produce products safely, and thus may increase the costs of meeting a particular level of safety or discourage innovation.

Two kinds of food safety regulation have come into wider use in recent years. The Hazard Analysis Critical Control Point (HACCP) system has been mandated for some part of the food system in the EU, Australia, New Zealand, Canada, and the U.S. The HACCP system specifies procedures to prevent food safety hazards (see box 2). It overcomes the high cost of information (testing for outcomes) that makes enforcement of outcome standards difficult for some hazards. Although HACCP appears to be a process standard, it is not as prescriptive as specifying GMPs. It encourages firms to focus on measurable indicators of hazards, and on the points in the process where hazards are most likely to occur. It allows firms some flexibility in designing and implementing controls to fit specific circumstances. Mandating HACCP also reduces the cost of regulatory enforcement. Rather than frequent inspection of product or of GMPs, a regulatory agency can review industry records periodically to verify that a HACCP plan is working.

Box 2 What is HACCP?

HACCP is widely recognized in the food industry as an effective approach to establishing good production, sanitation, and manufacturing practices that produce safe foods (Pierson and Corlett). HACCP systems establish process control through identifying points in the production process that are most critical to monitor and control. It provides feedback to direct corrective actions. HACCP's preventive focus is seen as more cost effective than testing a product and then destroying or reworking it (ICMSF). The system can be applied to control at any stage in the food system.

Seven principles are involved in developing and operating a HACCP program:

1. Assess the hazard, list the steps in the process where significant hazard can occur, and describe the prevention measures;
2. Determine critical control points(CCPs) in the process;
3. Establish critical limits for each CCP;
4. Establish procedures to monitor each CCP;
5. Establish corrective actions to be taken when monitoring indicates a deviation from the CCP limits;
6. Establish record keeping for the HACCP system;
7. Establish procedures to verify that the HACCP system is working correctly.

By focusing inspection at CCPs, HACCP improves the scientific basis for safety and control processes. A CCP is "any point in the chain of food production from raw materials to finished product where the loss of control could result in unacceptable food safety risk" (Pierson and Corlett). Monitoring of CCPs is done best by using indicators that can be measured easily. This focus on measurable indicators provides a more cost-effective approach to control than product sampling and testing, which is expensive and may not provide timely results. This is especially important for foodborne microbial pathogens, because their incidence is low and costs of testing are high.

Source: Unnevehr and Jensen (1999).

The second type of regulation coming into wider use is food product labeling to indicate potential hazards or methods of food production. This can allow consumers who are at higher risk to choose safer products or to prepare food so as to avoid risk. Thus it facilitates consumer choice and risk avoidance. It is also increasingly used to resolve controversies, such as those surrounding biotechnology. For example, the EU requires labeling of foods produced with GMOs.

How Quality Relates to Safety

Defining Quality and Safety

Food safety can be confused with quality. Quality includes all product attributes that influence a product's value to consumers. This can include positive attributes, such as flavor or pleasing appearance, and negative attributes, such as spoilage or safety hazards. If all quality attributes were detectable to consumers, then markets would determine the quality supplied.

The distinction between safety and quality has implications for public policy. Safety refers to hazards to human health in food. Quality refers to all attributes, and thus might include safety. Because market failures in safety are more common and public health is a strong justification for public intervention, it is often convenient to separate food safety from more general quality issues. Public efforts should focus on health hazards, and quality issues can be left to private industry management.

One reason why that distinction is difficult in practice is that many quality attributes are not detectable. So market failures may occur for quality attributes other than food safety. Industry might lobby for public quality certification in order to secure product reputation. This reputation may be considered a public good, and regulation is called for to prevent free riders from supplying bad quality in the market. This is the argument for a public role in setting quality grades and standards, and in certifying that products meet those grades and standards.

Quality grades and standards can be voluntary or mandatory. When they are mandatory, they can be a disguised means of limiting supply and increasing producer profits. Mandatory standards also have the drawback of limiting flexibility in meeting specific quality demands or in meeting new or changing quality demands. These are all arguments for limiting government's role in quality regulation to food safety alone. The government can mandate a minimum level of safety, and facilitate quality control for non-safety attributes.

Managing Quality and Safety

In practice, private management of quality and safety is increasingly complex and intertwined. There is a growing use of process control approaches from farm to table in food production for industrial-country markets. Such approaches, sometimes characterized as Total Quality Management, provide quality assurance that products will meet complex specifications. Internationally recognized certification, often through the ISO, is increasingly applied to food production and processing. The ISO provides a "standard for standards" through application of a framework for verifying the elements of a firm's production process that assure quality.

The management of safety is a subset of more general quality management (Mazzocco). The Hazard Analysis Critical Control Points (HACCP) system is used to address hazards that can be introduced at different points in the food chain or are difficult to measure. The HACCP system focuses on prevention and control of hazards, rather than on end-product testing (see box 2). An advantage of HACCP is to focus resources on the most important control points, which can minimize resources used to improve safety. Properly applied, HACCP may lead to process redesign, which can reduce the cost of providing quality. HACCP involves analysis of the entire system, with the corresponding need to coordinate preventive actions throughout the production process.

The management of quality and safety are linked in practice, and market failures can occur for many attributes, not just for safety. Thus, some combination of public and private actions may be needed to improve both food safety and overall food quality, particularly in export markets. Public intervention can provide leadership for improving quality and safety, and when well-designed, will not preclude private innovation.

Risk Perceptions and Cultural Approaches to Risk Management

Any risk management decision is made within the framework of competing interests and perceptions about food safety risk. This political economy has important consequences for agricultural trade and for domestic food safety regulation in client countries.

Disagreements at the Global Level

Who bears the risk of unsafe food or who bears the cost of risk reduction is frequently a highly political issue both within and between countries. This was reflected at the global level in the Agreement on Sanitary and Phytosanitary Measures under the 1994 GATT agreement (explained in the *Export Market Investments* chapter). The SPS agreement recognizes the value of harmonization (making standards the same across countries), but at the same time allows individual countries to choose their desired level of risk. Countries can choose very low risks, but standards must be based on scientific assessment of risk.

These issues remain very contentious, particularly between the EU, on the one hand, and the U.S., Canada, Australia, and New Zealand on the other hand. In the latter group of countries, quantitative risk assessment is often used as the basis for policy making. Consumers in those countries feel comfortable with the idea that risks can measured, and trade-offs can be evaluated to arrive at a reasonable level of acceptable risk. In the EU, there is less use of quantitative risk assessment, and more reliance on the "precautionary principle" (Jasonoff). This approach recognizes that some risks are unknown and cannot be measured, and therefore some precaution in avoiding them is wise.

Culture, Cuisine, and Outrage Factors

How consumers evaluate risk varies with the nature of the risk. Risks that are familiar, detectable, and naturally-occurring create less concern than risks that are unfamiliar, man-made, and undetectable.[2] This difference in risk perception may explain some of the paradoxes in risk tolerance observed around the world (Bureau and Marrette). The French tolerate a relatively high level of risk from foodborne pathogens in soft cheeses, for example, while they do not tolerate a relatively low, but less familiar, risk from genetically modified tomatoes. Americans tolerate relatively high risks from naturally occurring pathogens in undercooked hamburger, but insist upon only negligible risks from pesticide residues. These differences in perceptions about tolerable food risks will have parallels in developing countries, and will condition domestic policies about acceptable risks. Although microbial pathogens might be a larger quantitative risk, for example, public outrage may be greater for an incident of food adulteration, which violates trust and introduces an unfamiliar hazard.

---

[2] See *Understanding Risk Analysis* (http://www.rff.org/misc_docs/risk_book.htm) for a complete list of attributes that influence risk perception.

4. Export Market Investments

Fresh Food Product Exports from Developing Countries

Food and agricultural exports from developing countries account for about 10 percent of their total exports, which are primarily to high income countries. Fresh food products, which include exports of fresh meat, seafood, vegetables, and fruits, account for half of the value of total food and agricultural exports from all developing countries. In industrial countries, these fresh food products have a high income elasticity of demand, and in many cases have not been restrained by traditional agricultural trade barriers. Trade in these products has been expanding rapidly in high income markets, and the fresh product share of agricultural trade has increased in many developing regions (Thrupp).

Food safety issues are more likely to be a concern in fresh food product trade than in other kinds of agricultural product trade. First, fresh products are shipped and consumed in fresh form, so handling at all points of the food chain can influence food safety and quality. Manufactured or processed food products have more widely established and recognized standards, and may not deteriorate during shipping and handling. Second, standards in industrial-country markets are based on sanitation and good manufacturing practices that may not exist in less developed countries. Meeting such standards may require greater initial investments in quality control and sanitation in developing countries. Third, these fresh commodities are subject to increasing scrutiny and regulation in developed economies as food safety hazards are better understood and more often traced to their sources. Taken together, these fresh food marketing issues pose challenges for developing country exporters (see box 3 for two recent examples).

Table 2 U.S. FDA import detentions by product category, January-May 1999

Product Category	*Number of Detentions*
Vegetables/Vegetable Products	1,991
Fishery/Seafood Products	1,661
Fruits/Fruit Products	962
Candy w/o Choc/special/Chew gum	345
Spices, flavors and salts	212

Note: Table covers top five product categories.

Some evidence of the importance of SPS for developing countries exports and for fresh products is found in the U.S. FDA detention lists. These lists report imports detained for food safety or other technical violations. FAO (1999) summarizes the FDA data by reason for detention and region of the world for 1996/97. Most detentions were for contamination of food with insects and rodent filth, followed by microbiological contamination, and pesticide residue violations. Over half of rejections are attributable to lack of basic food hygiene and failure to meet labeling requirements (FAO 1999).

We examined the FDA detention data by product category for Jan-May 1999. Table 2 shows the total detentions by product category for the five most important categories. The top three categories are vegetables, fishery products, and fruits, confirming the importance of SPS issues for these fresh commodities. Meats had very few detentions, presumably due to the pre-certification of inspection systems in exporting countries carried out by USDA. We also examined the countries accounting for detentions. Detentions occur at a low frequency for many exporting countries, but those countries with the highest numbers of detentions for fresh commodities are developing countries.

Fresh Food Exports by Region and Destination Markets

Fresh food product markets would be the likely focus for most investments in food safety for export promotion. To understand how such investments might vary across regions, it is useful to examine fresh food product trade patterns by region and by destination market. The data for the following discussion are in the Appendix.

- East Asia has the largest total value of fresh food exports among the regions. Meat, fish and vegetables are all important. Most exports from East Asia go to Japan, with the EU and North American markets as significant, but much smaller, customers.

- The LAC region has the second largest value of fresh food exports. This region exports substantial amounts of all four fresh food product categories, with fruit being the largest. North America is the biggest market for all categories, but LAC countries also export to the European Community, and Japan is an important market for fish and meat exports.

- NA/ME exports are primarily fruit, fish, and vegetables. The EU is the principal market, and Japan is a secondary importer.

- Eastern Europe exports primarily meat and vegetables, almost all of which are exported to the EU.

- Sub-saharan Africa (SSA) exports go mainly to the European market, but some fish is exported to Japan. Most of the value of SSA exports is in fish and fruits.

- South Asia (SA) exports are primarily fish, but there has been rapid growth in meat and vegetables. South Asia exports to all three of the major high income markets.

Certain patterns are clear from these regional data. Relatively developed regions have a greater volume of exports. The LAC and EA regions are major participants in world food markets and have well developed market channels. They have much greater product and export market diversification than the other regions. The Middle East, North Africa, and Eastern European regions are more dependent on particular products and on the EU market. The regions with the greatest number of low income countries, SSA and SA, have exports that are more concentrated in particular products. Fish and seafood are particularly important for low income countries. In Africa, exports are concentrated in one market, the EU; South Asia has some diversity in export markets.

Box 3 SPS issues in fresh food exports from low income countries

These two cases show the importance of meeting sanitary and phytosanitary standards for developing new export markets. They also demonstrate the importance of managing quality and safety throughout the food production process and the public sector role in facilitating improved safety.

Snow Pea Exports from Guatemala

Non-traditional agricultural exports from Central America grew at 16% between 1983 and 1997. These exports, which are particularly important in Honduras, Guatemala, and Costa Rica, go to both North America and Europe. Case studies in Guatemala found that chemical overuse was the primary factor contributing to high detentions and rejection rates for shipments at ports-of-entry in the United States. During 1984-94 over 3000 Guatemalan shipments valued at over $18 million were detained and/or rejected at US ports of entry for chemical residue violations. Snow peas are an important non-traditional export from Guatemala, but insect and disease infestations have led to excessive reliance on chemical control measures. The 1995 leaf miner crisis resulted in a USDA Plant Protection Quarantine for all Guatemalan snow pea shipments. Research sponsored by the Government of Guatemala and USAID identified the leaf miner species as not exotic to the U.S, and therefore not a threat to US producers, and recommended control strategies to reduce chemical residues. Snow peas cultivated under integrated pest management (IPM) had reduced pesticide applications and lower product rejection rates. The plant protection quarantine (PPQ) was removed in 1997, re-establishing a $35 million annual market for Guatemala.

Frozen Shrimp and Prawns from Bangladesh

Bangladesh exports frozen shrimp and prawns to the EU, the US, and Japan, for a total value of $288 million in 1996. In July 1997, the EC banned imports of fishery products from Bangladesh. EC inspections of processing plants in Bangladesh determined that deficiencies in infrastructure and hygiene resulted in a potentially high risk to public health. In the US market, Bangladesh seafood was placed under automatic detention, so that each shipment is inspected and allowed to enter only if it meets safety standards. In 1997, 143 shipments of frozen shrimp from Bangladesh to the US were detained, usually for microbial contamination with *Salmonella*. In order to overcome these problems, both industry and government made major investments in more modern plants and laboratories and in personnel trained in HACCP procedures. HACCP was implemented as a nation-wide program in December 1997. This quality assurance legislation was recognized as equivalent to the EU, and the EU import ban was lifted for six approved establishments in December 1997.

Sources: Sullivan, et al. and Cato and Dos Santos.

These patterns suggest the kind of issues that will be important for export market development and for food safety issues in trade. Export market development and diversification are important for the low income countries. Continued export market access will be more important for the middle income regions. Eastern Europe is a special case because of the need to meet EU standards in order to integrate with the EU market. Exporters in all regions need to know how to meet standards in different markets and how to meet the increasing demand for product trace-back and certification of production methods. Lessons from successful Latin American and East Asian exporters may provide models for emerging exporters.

Both sanitary and phytosanitary measures will be important in fresh food markets. Pesticide residues and microbial contamination are issues for fruits and vegetables. Water quality, microbial contamination, and the need for a verified HACCP system in processing are issues for seafood and fish production. This product market is particularly important for low income countries. Microbial contamination, drug residues, and inspection systems are issues for meat exports. Exporters would benefit from recognition of equivalence across high income markets and from the use of transparent product standards.

Meeting Quality and Safety Standards in Export Markets

There are many different models for vertical coordination to meet quality and safety standards. Production in an developing countries might be tightly controlled by a multi-national firm for export to a high income market. The production process might be wholly owned or contracted with local growers. Alternatively, local growers may be coordinated through an exporting firm that provides guidance on quality standards and assurances to importers. The public sector in developing countries can play in role in testing and certifying export quality. The public sector in the importing country may facilitate trade by providing pre-certification for exports through in-country inspections of the production processes. This can substantially reduce costs associated with detentions and rejections.

Jaffee reviewed success stories in exporting high-value food commodities for the Bank in 1992. He found that successful exporters were able to compete on quality and product differentiation, not just as a source of low-cost supply. The private sector played the dominant or exclusive role in developing the export market and a high degree of vertical integration was an important feature of successful export production. Often foreign direct investment played a role, but was not the dominant force. Government assistance was important in providing necessary infrastructure, certification that standards were met, market information about standards in importing countries, and research to improve production methods. These elements are present in some Bank projects that focus on export promotion, as will be discussed in chapter six.

Special Issues Facing Small Countries

The Rural Development mission area has a special commitment to poverty alleviation in the 41 low income food insecure countries, many of which are in sub-Saharan Africa. Do high SPS standards in developed markets reduce the ability of these countries to export agricultural products or to produce domestic food at a cost competitive with imports? Whether SPS reduces the ability of small countries to compete in world agricultural trade depends on the scale of production required to support quality and safety management. There are examples in the literature and from Bank project experience of non-traditional agricultural exports from small countries. Export quality control is labor and management skill intensive. With the right investments in management, small countries can compete in export markets. Because there are strong private incentives for these investments, the scale of public activity is not an issue for exports. However, one issue that deserves further exploration is how increased quality and safety management influences returns to labor or to small agricultural producers (Thrupp).

On the import side, the small low income countries' lack of competitiveness and food insecurity is the result of lagging agricultural productivity in producing the major food staples. Yields of major grain crops in the food insecure countries are much lower than elsewhere. This lower productivity is not related to food safety or international sanitary standards. Competing with grain exports from industrial countries will require increased productivity, not meeting higher SPS standards.

The SPS Agreement and the Framework for International Trade

While exporters must meet the quality and safety demanded by import market consumers, there are international trading rules to ensure that public standards are applied fairly and equally to both domestic and imported products. When an exporter is denied access due to sanitary or phytosanitary standards, there is recourse for WTO members under the WTO dispute process if the action appears to be unfounded or unfair. The 1994 GATT agreement included the Agreement on the Application of Sanitary and Phytosanitary Measures (SPS agreement).

The SPS agreement covers trade measures that protect human, animal, or plant life or health. It sets out several ground rules for such measures, with the intent of ensuring that they do not pose unfair barriers to trade. These "ground rules" include:

- Transparency—Nations are required to publish their regulations and provide a mechanism for answering questions from trading partners.

- Equivalence—Member nations must accept that SPS measures of another country are equivalent if they result in the same level of public-health protection, even if the measures themselves differ. The same level of health protection should apply to both domestic and imported products.

- Science-based measures—Regulations cannot impose requirements that do not have a scientific basis for reducing risk.

- Regionalization—The concept of pest- or disease-free areas within an exporting country is recognized. Exports can be allowed from such areas, even if other areas of an exporting country still have the disease or pest.

- Harmonization—Member nations recognize the desirability of common SPS measures. Three international organizations are recognized sources of internationally agreed-upon standards: the Codex Alimentarius Commission, the International Office of Epizootics, and the International Plant Protection Convention.

- National sovereignty—Countries may choose a risk standard that differs from the international standard. This recognizes that individual nations are unwilling to subscribe to uniform international standards for all hazards.

- Dispute Resolution—There is a clearly defined mechanism for resolving disputes between countries in a timely manner. The dispute settlement panel is expected only to state whether the SPS measures under question have a scientific basis and are consistently applied.

During 1995–97, there were seven formal complaints under the SPS agreement. Most of these complaints involved disputes between industrial countries, but three complaints were against Korea by the US, EC, and Canada. Hungary and Brazil joined the US and the EU in a complaint against Japan's varietal testing requirements. Some disputes were resolved through negotiation. Only one dispute, a complaint from the US and other meat exporters (Canada, Australia, New Zealand, and Norway) regarding the EU ban on growth hormones in meat, has reached the stage of a final report. The dispute panel decided that this measure did not have a scientific basis, but this finding is still the subject of controversy between the U.S. and the EU.

The existence of the SPS agreement has provided a broader catalyst for regulatory reform (Roberts). This agreement has prompted actions to open markets. The U.S. recognition of disease free zones for Argentine beef and for Mexican avocados is an example of implementation of the principles of the SPS agreement. Another example is the settlement of the three complaints against Korea through formal negotiations under the SPS dispute process. Worldwide efforts to better document SPS measures is a response to the agreement's mandate for greater transparency.

Optimism about reducing SPS barriers might be premature, however, given the current disputes between the U.S. and EU over food risks and risk management. Tensions between the two trading partners now focus on guarantees regarding use of risky inputs. As this is written, the EU is trying to find ways to comply with the beef growth hormone decision without allowing beef imports. One remaining issue is whether beef exporters can guarantee that growth hormones are always properly used. This controversy highlights the importance of guaranteeing safe production practices from farm to table. Another controversy concerns the acceptability of genetically modified food crops.

While this controversy relates only marginally to food safety, it demonstrates that different approaches to regulating biosafety will be difficult to resolve. These controversies create uncertainties for Bank client countries about the acceptability of production methods and products across potential markets.

Potential Areas for Developing country Focus in WTO 2000

How to regulate trade of potentially risky food and agricultural products will be a contentious issue in WTO 2000. The current disagreements regarding risk management at the international level could lead to a re-examination of the SPS agreement. This could bring about new trade barriers that might hinder growth and development for developing countries. Resolving these disagreements in a way that will not preclude trade expansion is a serious challenge for all of the participants in WTO 2000. Developing countries will need to evaluate their own interests in terms of trade opportunities and domestic risk aversion in order to participate fully in this debate. While GMOs and biotechnology will be in the spotlight, the more mundane sanitary and phytosanitary issues relating to long-established risks are also important for developing countries.

If the framework of the existing SPS agreement remains intact, then there are three specific areas for developing countries to consider. The first is the role of the Codex, the OIE, and the IPPC in setting internationally recognized standards. Many developing countries do not have the capacity to participate in these international organizations and it is not clear that their interests are always taken into account. Learning how to participate effectively involves capacity building within the public sector in developing countries and may also involve forming coalitions around issues of mutual regional interest. FAO has provided some assistance to developing countries on this issue and the Codex has regional committees, but further capacity building is needed.

A second issue is the growing use of process standards for food safety and the resulting difficulties in determining equivalence across countries. Equivalency in risk outcomes is the de jure standard under the SPS agreement. However, in practice, equivalency might be determined through requiring particular processes from particular countries. Whether risks really differ for commodities produced in developing countries must be examined carefully on a case by case basis. Developing countries need to participate in the current discussions regarding risk assessment and determination of equivalency that are proceeding within the Codex to ensure that they are not held to a de facto higher standard.

The third issue is whether a domestic food safety regulatory system is becoming a necessary prerequisite for participation in trade. Some components of a food safety system may be necessary, but developing countries should resist the trend in importing countries towards requiring equivalent systems rather than equivalent outcomes for specific products (see Bangladesh example in box 3). Some policy analysis and risk assessment capacity is necessary for participating in international trade negotiations. Some control over animal and plant hazards, or monitoring of imports to prevent hazards may also be necessary to ensure trade participation. It is possible that such capacity can be developed at a regional level, if there are common trade interests and similar risks. A full scale legal framework for food safety regulation with minimum standards is expensive for small low income countries, and may be counterproductive in any country without sufficient public capacity for enforcement. In the next section, we discuss further the interactions between trade and domestic regulation.

5. Investments to Improve Food Safety in the Domestic Food Market

Interactions with Export Markets

Will food safety investments to meet export market standards also improve domestic food safety? Positive spillovers for domestic markets can be expected from investments in infrastructure or regulations for processing hygiene or increased control of animal/plant health hazards with human health implications. Seafood products would seem to be an industry where hygiene improvements for export would have benefits for local consumers. Jaffee in his review of developing country food export success stories found that there was often a strong interaction with the domestic market. Domestic demand for a product provided producers with experience and an assured fallback market. Often, however, the domestic market consumed a lower quality product (although not necessarily less safe). Alternatively, certain kinds of export production may be enclaves with higher quality and safety to meet world market standards. There are examples from the literature (see Guatemala in box 3) and from Bank experience of export food production with little or no domestic market or of two-tier markets with one production process for exports and another for the domestic market (e.g. Argentine beef).

Is the latter outcome undesirable? New export markets provide income generation and may be expected to improve health and well-being primarily through increasing household income. Furthermore, the relative importance of food safety risks and the market mechanisms for determining who bears risk will differ between most developing countries and most industrial countries. Thus it may or may not be beneficial for export standards to apply to domestic production, even when the commodity is widely consumed locally.

How Markets for Food and Food Safety Change During Development

The responsibility for food safety shifts from the consumer to the producer during development, just as many other activities shift from the household to the market (e.g., food production, water supply). Consumers bear most food safety risks and must take actions to mitigate those risks, either through food preparation or choices among food sources. As markets develop, and more food preparation is undertaken outside the household, then responsibility shifts to producers. A formal sector emerges alongside the informal sector. A processed food product may be available both from informal vendors and also from a large private firm with a brand name. Consumers begin to demand better food safety through their actions in the marketplace and through the political process. Because food safety is often a public good, governments intervene to regulate production practices in the formal sector. The issue for Bank client countries is at what point in the development of the food system does intervention make sense, and what kinds of interventions make sense.

Figure 1 Public activities to improve food safety

Type of Public Activity	How Activities Evolve as a Country Develops			Project Components to Support Public Activities
	Low Income	*Middle Income*	*High Income*	
Policy Decision-Making Capacity	Stakeholder involvement in policy making Disease or Hazard Surveillance Participation in Codex, OIE, IPPC			Legal and regulatory framework Infrastructure to support disease surveillance or monitoring capability Risk assessment training Participation in international organizations
	Qualitative Risk Assessment to inform risk management	Quantitative Risk Assessment and Cost-Benefit analysis		
	Adapt standards from Codex; or major importer for niche markets			
		Set standards according to local risk conditions and preferences		
Provision of Information	Targeted interventions for reduction of childhood illness and malnutrition	Consumer and Industry Education for better food handling and preparation	Labeling and certification to inform consumers about production processes, product safety, and potential hazards	Market information about import standards Voluntary certification system Training programs for producers and consumers
Measures for Prevention and Control	Hygiene, training at key points in food chain	Control of external or single source hazards Phased imposition of standards Monitoring of key hazards in food supply	Mandated standards Widespread application of HACCP Monitoring of food supply	Control programs for single source hazards Phased imposition of regulation for formal food sector Provide generic HACCP models for small processors and food vendors
Infrastructure and research	Water Supply Sanitation Marketing facilities Applied research to reduce key hazards		Basic and applied research on many hazards	Sanitation and water supply Marketing infrastructure Research to develop hazard control

Source: Laurian Unnevehr, presentation at IICA/EMBRAPA/World Bank workshop on "The Future of Food Safety", August 26-27, 1999.

Figure 1 shows the types of public activities that can improve food safety, how these activities evolve during the process of development, and examples of related project components for investment. Food safety interventions build from basic investments and simple interventions to more complex regulatory systems as economies develop. Priorities for public action change at different levels of development. At low levels of income, investments in water and sanitation and targeted interventions to reduce child malnutrition are the highest priority for food safety. As food systems and the capacity for food safety policy develop, then targeted interventions for single source hazards or important control points in the food marketing chain become practical. At higher levels of income, more extensive regulation and enforcement are feasible. These would include setting specific standards for food products (e.g. pesticide residues) and monitoring the food supply to ensure compliance with these standards.

In low income countries, food safety is closely linked with sanitation, water supply, and nutrition. Because the burden of illness, particularly for children, is a function of all of these variables, it is difficult to address food safety risks separately. Does a food safety hazard in a child's food arise from the environment, food preparation, water sources, or food production? Sometimes it may be from all of these potential sources. Thus removing one source of hazard may or may not reduce the consumer's risk. In very poor countries or households, the provision of basic water and sanitation services may be a prerequisite for efforts to reduce the most important food safety risks. However, such investments are not always sufficient for achieving improved food safety. Consumer and food handler education is an essential complement to any investments in sanitation (Moy et al). Thus projects to improve health through sanitation and water could also address the interface with food safety. Box 4 provides some examples from Kenya.

At low levels of development, countries will likely adapt internationally recognized standards from the Codex Alimentarius as needed, because domestic capacity for risk assessment will be limited. As countries develop, they will gain greater public capacity for foodborne disease surveillance, risk assessment, and participation in international organizations. These can provide the basis for setting standards and designing interventions according to local risk conditions and preferences, and for educational efforts aimed at consumers and industry. Identification of the most important risks can lead to public research to reduce specific hazards, such as mycotoxins in food staples.

Decisions about public interventions to prevent or control food hazards must recognize the limited public capacity for enforcement and the need for prerequisite investments in food marketing infrastructure. As food systems develop, urban markets, street food vendors, and slaughterhouses are highly visible points of concentration in the food chain and are therefore often the focus of initial interventions to impose standards of hygiene. In the Philippines and in India, for example, there are local government monopolies over small slaughterhouses that serve the informal market, to ensure a minimum level of hygiene. The WHO reports that street food vendors are regulated in three-quarters of their member nations, with measures such as registration, periodic medical exams, or periodic required training.

Box 4   Food safety is one part of targeted interventions to reduce child malnutrition: examples from the Kenya poverty assessment

The World Bank's Kenya poverty assessment identified targeted interventions to reduce child malnutrition. The assessment and suggested interventions implicitly recognized the role of food safety in improving children's nutritional status and food security. Variations in child malnutrition among poor households were explained by freshness and cleanliness of food, the time and care taken in food preparation, the availability of water and sanitation services, the mother's health status and nutrition knowledge, and other variables such as age and sex of the child. The suggested interventions included mother's education, immunization, nutrition education on weaning practices, including hygienic preparation, provision of water and sanitation, and enhancing the mother's nutritional status. This assessment shows how food safety is one component of many linked factors determining child malnutrition. By addressing food preparation, cleanliness, water and sanitation, the interventions would improve food safety and reduce the incidence of children's diarrheal disease.

Source: World Bank, Kenya Poverty Assessment, 1995.

Such measures can improve food safety but they could also raise food costs or drive the informal food sector underground. One of the difficulties that developing country governments face is how to identify enforceable interventions as the formal food sector grows, without driving out informal activities that still serve an important economic function. Some guidelines are possible for the evolution of domestic market controls. First, the control of single source hazards is the type of intervention that is more easily implemented when public capacity is limited. Some examples are hazards that enter only through international trade or hazards that have only one source, such as a parasite or disease carried only by one animal species. Interventions can be designed to address these kinds of hazards more easily than interventions for elimination of hazards that persist in the environment and can enter food at many points. Of course, whether any particular intervention is warranted would have to be examined through a cost benefit analysis, and interventions would have to be designed to create incentives for eliminating the hazard rather than circumventing controls.

Second, limited public capacity in less developed countries suggests that an emphasis on sanitation investments, risk prioritization, training for industry and food handlers, and provision of information to consumers is the feasible approach to food safety improvement. These kinds of activities help the entire food system without penalizing the informal sector, and accord with the principles emerging from food safety regulatory reform in developed economies.

Special Issues Facing Small Low Income Countries

Small low income countries may have higher costs per capita in regulating and monitoring food safety. These countries have limited public capacity and underdeveloped food systems. Thus, implementation of many food safety interventions is premature. In small, low income countries, water, sanitation and targeted interventions to reduce diarrheal disease in small children would have higher priority than comprehensive food safety regulation. The Bank publication, *Can the Environment*

Wait?, argues that investing in water and sanitation is an immediate priority in East Asia, and summarizes the benefits from making such investments early in the development process.

As a first step towards establishing food safety regulation within small countries, certain elements of regulation may be established through regional efforts. For example, countries within a region might adopt the same standards or use the same framework to monitor foodborne disease. These regional efforts might facilitate the necessary activities for effective participation in international trade negotiations and in international standard setting agencies. The efforts of PAHO in LAC and of ASEAN in Southeast Asia may provide some useful models and lessons.

New Models for Food Safety Regulatory Systems

Food safety regulation has undergone significant change in many developed economies during the last decade. This evolution of regulatory models provides some lessons for countries now building their own public systems.

It is very common for regulatory responsibility in food safety to be fragmented across several different government entities operating under different laws, with resulting overlaps and gaps. Fragmentation comes about because regulation of different hazards and different commodities evolves in a piecemeal fashion with new scientific evidence, with public outrage over a particular incident, or with corollary regulatory activities for a particular industry. Fragmentation may be less efficient and effective, because resources are scattered and cannot be shifted towards the most important sources of risk as they change.

At least 16 countries are considering some restructuring of food safety regulation, and four countries are in the process of consolidating their food safety regulation under one agency (GAO). An important issue in restructuring is how the new agency is placed in relationship to other public responsibilities. Two countries in the process of consolidating food safety regulation, Ireland and United Kingdom, will have their new agencies report to a minister of health. The other two countries, Denmark and Canada, have agencies that are responsible to agriculture and food. These choices reflect the tension between health and agricultural interests in the regulation of food safety.

A recent United States General Accounting Office report summarized lessons shared by these four countries from their restructuring experiences. All countries expect benefits in terms of more effective public performance in the long run, including improved efficiency, greater ability to provide farm to table oversight for the whole food system, and enhanced international market access. The new agencies expect to have the ability to shift resources over time to areas of greatest risk.

Along with consolidation, there is also a general move towards regulation that is less prescriptive and that shifts responsibility for food safety from government oversight to the food industry. The new food safety standards for Australia are one example of these trends, which can also be observed in Canada, New Zealand, and to a lesser extent in the

U.S. and the EU. Rather than inspect products or processes, regulatory agencies find it more efficient to require industry to adopt preventive measures under a HACCP system (see *Public Interventions to Improve Food Safety* section). This approach can also include training for industry and for food handlers in HACCP techniques and good hygiene. The key elements of this approach are prevention, flexibility, training, and shared responsibility among industry, government, and consumers. The FAO recognizes these elements as important for action in developing countries.

To sum up, current understanding of food safety management and the desire of most industrial countries to be responsive to consumers and efficient in the use of public resources has brought about changes in food safety regulatory systems. A progressive food safety regulatory system includes:

- Consolidated authority with ability to address the food system from farm to table and to move resources towards the most important sources of risk.

- Use of comparative risk assessment to help prioritize public action.

- Cooperation with industry and consumers to provide information and education.

- Use of HACCP principles to promote prevention and industry responsibility in place of prescription and inspection.

- An open decision making process that allows stakeholder participation.

- Evaluation of public health outcomes.

6. Bank Project Experience Related to Food Safety

Overview of Bank Projects Related to Food Safety

The ESSD internal web site contains a database of 65 projects with elements related to food safety, quality control, or animal and plant health (see table 3). For brevity, we use the term "agricultural safety" below to refer to issues of food safety, animal health, and plant health. There are many similar projects in the areas of animal and plant health. The identified projects with quality control or food safety aspects are fewer and more diverse.

The components of these projects are summarized in table 4. The similarities among these components show the overlap among agricultural safety issues in terms of public services and client country needs. Generally the project components fall into one of three categories: *physical capital investments, institutional or regulatory capacity building,* and *public support services.*

- Physical capital investments include new equipment for public laboratories engaged in product testing or disease surveillance, investments by private firms in processing, storage, or other marketing capacity, and public investments in infrastructure to facilitate marketing or improve quality such as cold storage facilities at an airport.

- Institutional or regulatory capacity building includes establishing the legal framework for border SPS measures such as inspection or quarantine, establishing grades or standards for agricultural products or inputs, and strengthening public capacity to set policy in these areas.

- Public services to support private production include animal and plant disease eradication, or market information about export opportunities and standards.

Many of the projects have a strong export orientation or motivation. The need to establish public services or a legal framework is often justified for export market development, even if the project does not have a specific export promotion goal. For example, projects in Eastern Europe frequently cite the need to emulate EU standards to facilitate full participation in EU markets. Many projects in sub-Saharan Africa focus on non-traditional exports, including livestock and meat. In contrast, processing and marketing investments for domestic markets were cited as a motivation more often in large countries such as China and India, where the size of the market may justify greater public investment for domestic consumers.

Surprisingly, there were relatively few projects in the highly successful exporting regions of LAC and East Asia. Two exceptions are the highly successful project in Brazil to eliminate FMD in order to facilitate beef exports (Coirolo) and the Argentina project to strengthen plant and animal disease regulation. Does the small number of projects reflect low demand or need in these countries, or a lack of Bank participation in meeting needs? If the latter, it may reflect Bank and client country priorities for Bank funded projects.

For many projects, agricultural safety issues are a very minor part of a much larger project for the entire agricultural sector. The required loan amounts to support the project components in table 4 are not large, and thus do not always justify a separate project. For several projects, the marketing or quality control components were revised or dropped in the process of implementation. Larger goals, investments, or policy reforms took precedence over these safety concerns. Agricultural safety concerns are more likely to be addressed thoroughly within the context of projects oriented specifically towards export promotion, animal or plant health, or domestic food regulation.

These project components are generally recognized as areas where the public sector has an important role to play. It is interesting to consider the interaction of these components with policy reforms in agricultural markets and the partnerships between the public and private sectors. One major policy reform in several animal health projects is the privatization of veterinary services or the imposition of cost recovery for public veterinary services (see Umali, Feder, and de Haan for more discussion). This follows general Bank policies favoring privatization. The principles developed by Umali et al. are based on the structure of animal agriculture, and might be adapted to the structure of food processing as a guide for evaluating privatization of quality control.

Apart from the privatization of veterinary services, there are few policy reforms in these projects that specifically relate to agricultural safety or quality control goals. The exceptions are small and focused reforms, such as deregulation of transportation monopolies that hinder trade. Policy reforms may not be necessary for non-traditional agricultural exports, because these products have few existing policies that hinder trade. As a group, all of the projects in the database tended to strengthen the public role in agricultural market regulation rather than reduce it. This is appropriate given the weak institutions for dealing with agricultural safety issues in most client countries.

One area where the private sector role is more apparent is in quality control. Two interesting examples are found in projects for Senegal and the Cote d' Ivoire. These projects provided technical assistance to facilitate private producer organizations. These institutions took responsibility for identifying export markets and product specifications in those markets, and for establishing farm to export quality control, including traceback mechanisms to the farm level. The public sector role was limited to provision of phytosanitary certificates, but other aspects of quality and safety were certified by the private sector. This provides an example of Bank project facilitation of private sector quality control for non-traditional exports.

The quality control projects provide important lessons for food safety, because they often involve some element of institutional change. This non-quantifiable element of enforcement, compliance, and general awareness within the food system is the most important part of improving food safety.

We included three projects with food fortification activities in the database, because it was expected that this would be tied to quality control. However, these projects generally relied on provision of fortification through specific feeding programs, rather than the general food supply. This speaks to the need for broader food regulation as a prerequisite for food fortification.

There are very few projects with any true food safety component in the database, and of these, two have been dropped as Bank projects. Some export promotion/quality control projects include a small sanitary component: two fishery projects in China and Tunisia, and a Philippine coconut project. There are three Eastern Europe projects (Estonia, Albania, and Hungary) that provided support for domestic laboratory facilities to strengthen public food quality control within the context of larger agricultural sector investments. Outcomes were measured in terms of increased public sector activity (e.g., number of laboratory tests); the linkages to food safety risk reduction or to health outcomes were not clearly articulated.

Two projects that focused specifically on food safety regulation have been dropped. One for agricultural standards in Hungary was replaced by grant assistance from the EU. This example raises the issue of sources of financing for food safety investments. The other food safety project was developed for India and appears to be unique in the Bank portfolio. It proposed a comprehensive overhaul of food regulation under the Ministry of Health, in conjunction with reform of drug regulation. The two key components were increased laboratory testing to monitor for food adulteration (in response to consumer outcry) and phased implementation of HACCP for the food processing industry. The project development included discussions with consumer NGOs. This project was shelved when the government of India changed its priorities.

Lessons for Future Food Safety Investments

The successful animal and plant health projects provide some examples of components that could be supported to improve food safety, including laboratory tests to monitor the food supply, mobilizing industry support for public efforts, and cost recovery for public activities that benefit the agricultural industry. But the goals of animal and plant health projects, and the constituencies, are fundamentally different from food safety. Animal and plant health increase productive capacity, and industry will therefore support public actions that cannot be undertaken by individual producers. Food safety is a public health goal that benefits consumers and may impose costs on producers. Therefore agencies with experience in animal and plant health issues may have some of the right technical expertise for food safety, but will not necessarily be oriented towards measuring the right outcomes. A consumer and public health orientation in institutions and in policy will be necessary for achieving improvements in food safety.

The export promotion projects provide the clearest lessons. These include the importance of partnerships between the public sector and the private sector, the importance of training and market information, the close connection between safety and quality management in practice, and the need for small, but crucial, infrastructure investments or policy reforms. These projects and other studies of successful non-traditional food exports suggest that the best practice is to rely on private certification and private investments as much as possible. The strong market incentives in exporting should motivate both.

The small number of projects for domestic food safety regulation raise some difficult issues. The first is how such regulation will interact with other policy reforms. In *Reforming Agriculture: The World Bank Goes to Market*, this tension appears in discussion of regulatory and legal reforms. On page 85, this review of AGSECALs states: "Residual market controls applied in the name of public objectives ... serve to protect established interests, restrict the activities of certain ethnic groups, and supplement the salaries of public officials." Then on page 88, the report encourages legal and regulatory reform: "Some regulatory and legal reforms are vital for agricultural adjustment... Examples include ... phytosanitary procedures." This speaks to the tension between fostering a weak and potentially corrupt public role versus strengthening public capacity for an appropriate public activity. Resolving this tension will only occur on a case by case basis as opportunities to strengthen institutions and change regulatory approaches are identified.

The second difficult issue is how to relate project components to the overall elements of a progressive food safety regulatory system discussed in Section Four. Investments in laboratories alone do not bring about a comprehensive and effective system. At the very least such activities should be linked to enforcement and to measured outcomes in food safety improvement. But more importantly, such investments need to be part of a larger effort that includes the legal framework for regulation, capacity building for comparative risk assessment and for HACCP approaches, and institution building for greater stakeholder involvement. Ideally, a comprehensive approach would also identify key infrastructure investments to support food safety through better sanitation. Box 5 provides some questions to guide a project appraisal for domestic food safety regulation.

Recommendations for Bank Action

There is scope for greater Bank leadership and investment in food safety. Leadership activities include fostering dialogue about SPS issues at the international level and fostering regional partnerships to address issues of mutual interest. Specific project investments are likely to be part of investments in public health, regulation of agricultural safety, or export promotion. Although food safety stand-alone projects will likely not be large enough to warrant Bank involvement, the Bank can broaden the scope of its food safety investments and implement the best practices identified above. In designing and implementing projects, the Bank can draw upon many other institutions for expertise in food safety, including the FAO, bilateral assistance from the major industrial country importers, and various private sector organizations (see chapter 7). The Bank can ensure that food safety investments are undertaken within the appropriate policy framework so that food markets can develop most efficiently.

Box 5. Questions for understanding a domestic food safety system

What is the current legal framework authorizing food safety regulation? Which kinds of hazards are covered? Which government agencies are involved? Is there legal liability for the food industry under some circumstances? How are food safety regulations and laws enforced? Answers to these questions will identify gaps in the legal and regulatory framework for food safety.

Which government agencies are involved in regulating food safety? What are their budgets and staff? Do they report to ministries of health, agriculture, trade, other? What is their training and background? If hazards are monitored, what use is made of the information? Answers to these questions will show the current public capacity for food safety policy and enforcement.

Who are the important stakeholders representing the food industry and food consumers? What issues do these groups think are most important? Answers to these questions will identify the food safety interventions that would have broad public support.

What are the most important food safety hazards and what are their sources? What are the most important foods in the diet and what kind of risks do they carry? What are the most important diseases and which ones can potentially arise from contaminated food? Which hazards can be controlled by the consumer or food preparer? Which ones can only be controlled during the production process? What proportion of the food supply is marketed to urban areas? Which food products are likely to pose the greatest hazards to urban consumers? Answers to these questions will help to identify targets for initial interventions, by identifying the most important risks that can be addressed during food production and processing.

What kinds of quality and safety control do large scale food companies follow in their processing? What kinds of quality and safety control are followed in the tourist industry? Do these control programs suggest areas for public investment to remove constraints? Do they suggest model HACCP plans that could be adapted for training processing firms in the informal sector? What existing programs could provide vehicles for training food preparers, such as food and nutrition intervention programs or licensing of street food vendors? Answers to these questions will show how the private sector copes with food safety, and whether private models could be adapted to interventions for the informal food sector.

Source: Unnevehr, Laurian and Hirschhorn, Nancy. 2000. Designing Effective Food Safety Interventions in Developing Countries in Giovannucci, Daniele, ed. 2000. The Guide to Developing Agricultural Markets and Agro-Enterprises.
World Bank home page: http://wbln0018.worldbank.org/essd/essd.nsf/Agroenterprise/agro_guide

---

Table 3 Agricultural health, food safety and quality control projects

Country	Fiscal Year	Project Name	Loan Amount (US$ million)	Safety Component
* Food Safety				
Albania	1996	Agro-Processing Development Project	6	Quality assurance, increase hygiene, quality control laboratories.
China	1998	Sustainable Coastal Resource Development Project	100	Seafood product quality and safety, includes HACCP training, adoption
Estonia	1996	Agriculture Project	15.3	Food quality control & veterinary laboratory, bacteria counts in dairy products
Hungary	1988	Agroprocessing Modernization Project	70	Grading, standardizing, quality control, testing products for residues.
Morocco	1999	Pilot Fisheries Development Project	5.2	Quality assurance on-board vessels, spread of good hygienic practice
Philippines	1990	Small Coconut Farms Development Project	121.8	Quality management, laboratories, survey of aflatoxin and incidence.
Tunisia	1994	Agricultural Sector Investment Loan Project	120	Quality standards, disease suiveillance, national vaccination, disease lab.
*Plant Health				
Algeria	1989	Desert Locust Control Project	58	Locust control, surveillance, plant protection, pesticides handling
Algeria	1994	Desert Locust Control Project	30	Locust control, surveillance, plant protection, pesticides handling
Argentina	2002	Prov. Agriculture Development 2	125	Animal disease control and plant protection
Argentina	1991	Agricultural Services & Institutional Development Project	33.5	Laboratory, certification, grading, classification, quality control, registration
Bangladesh	1991	Agricultural Support Services Project	35	Training and extension on improved seeds as well as seed certification
Brazil	1999	Animal & Plant Health Protection Project	44	Animal and plant disease control

Table 3 Agricultural health, food safety and quality control projects

Country	Fiscal Year	Project Name	Loan Amount (US$ million)	Safety Component
Cambodia	1997	Agriculture Productivity Improvement Project	27	Integrated pest management, disease control.
China	1990	Hebei Agricultural Development Project	150	Quality control and management
China	1991	Mid-Yangtze Agricultural Development Project	64	Improve fruit quality through grading and usage of disease-free seedlings
China	1996	Seed Sector Commercialization Project	100	Lab, external quality control and variety evaluation, seed certification
Ethiopia	1995	Seed Systems Development Project	22	Quality control and labeling of seeds, seed inspection and certification
Guinea	1988	National Seeds Project	9	Production of high quality seed
Kenya	2000	Agriculture Sector Investement	27	Disease control
Kyrgyz Repub	1998	Agriculture Support Services	15	Crop protection/plant quarantine and quality seed control
Mauritius	1991	Environmental Monitoring & Development Project	12.4	Integrated pest management
Moldova	1996	First Agriculture Project	10	Vinestock certification, improve quality and integrated pest management
Morocco	1994	Second Agricultural Sector Investment Loan	121	Plant protection
Sudan	1989	Southern Kassala Agriculture Project	20	Improve crop quality, plant protection
Ukraine	1995	Seed Development Project	32	Standard laboratories, seed quality inspection and quarantine inspection
Uzbekistan	1995	Cotton Sub-Sector Improvement Project	66	Grading, seed certification and seed quality control
Animal Health				
Argentina	2002	Prov. Agriculture Development 2	125	Animal disease control and plant protection

Table 3 Agricultural health, food safety and quality control projects

Country	Fiscal Year	Project Name	Loan Amount (US$ million)	Safety Component
Argentina	1991	Agricultural Services & Institutional Development Project	33.5	Laboratory, certfication, grading, classification, quality control, registration
Brazil	1987	Livestock Disease Control Project	51	Foot-and-month disease control, animal disease laboratories, quarantine
Brazil	1999	Animal & Plant Health Protection Project	44	Animal and plant disease control
Burkina Faso	1989	Agricultural Support Services Project	42	Strengthening and reorientation of preventive and curative animal health
Burkina Faso	1998	Agricultural Services II	41.3	Veterinary services, training and extension.
Central African Republic	1995	Livestock Development and Rangeland Management Program	16.6	Training for veterinarians and sales of veterinary pharmaceutical
Cambodia	1997	Agriculture Productivity Improvement Project	27	Integrated pest management, disease control
Cameroon	1989	Livestock Sector Development Project	34.6	Livestock disease control, tse tse fly eradication
Chad	1988	National Livestock Project	18.6	Distribution of veterinary pharmaceuticals and animal vaccination
Croatia	1996	Farmer Support Services Project	17	Veterinary services and epidemiological services
Ethiopia	1987	Fourth Livestock Development Project	39	Animal health services, drug supply and animal disease control
Ghana	1993	National Livestock Services Project	22.5	Acess to livestock health services & animal disease control
Ghana	2000	Agricultural Services	40	Animal disease control
Guinea	1996	National Agricultural Services Project	35	Increased vaccination coverage
India	1988	Second National Dairy Project	360	Veterinary services, foot and mouth disease control
Indonesia	1986	Smallholder Cattle Development Project	32	Veterinary services, disease investigation and disease eradication

Table 3 Agricultural health, food safety and quality control projects

Country	Fiscal Year	Project Name	Loan Amount (US$ million)	Safety Component
Kenya	1987	Animal Health Services Project	15	Disease protection, control surveillance & veterinary laboratories.
Kenya	1993	Emergency Drought Recovery Project	20	Trypanoci drugs, clinical drugs, vaccination and animal treatment.
Kenya	2000	Agriculture Sector Investment Project	27	Animal disease control
Macedonia, F	1996	Private Farmer Support	7.9	Animal health border inspection, veterinary services
Madagascar	1991	Livestock Sector Project	19.8	Disease control, training, tightening border sanitary controls, vaccination
Somalia	1979	Central Rangelands Development Project	8	Training, extension, veterinary services
Somalia	1989	Central Rangelands Research and Development Project, Phase II	19	Livestock disease eradication, training and extension
Somalia	1986	Livestock Health Services Project	19	Livestock disease control
Uganda	1991	Livestock Services	21	National disease control, tsetse control, veterinary privatization
Zambia	1995	Agricultural Sector Investment Program	60	Improve disease prevention and control
Quality Control				
Albania	1996	Agro-Processing Development	6	Quality assurance, hygiene and quality, quality control laboratories.
Argentina	1991	Agricultural Services and Institutional Development	33.5	Laboratory, certfication, grading, classification, quality control, registration
China	1990	Hebei Agricultural Development	150	Quality control and management
China	1991	Mid-Yangtze Agricultural Development	64	Improve fruit quality through grading and usage of disease-free seedlings
China	1992	Guangdong Agricultural Development	162	Improve quality of domestically marketed and exported agric. Products

Table 3 Agricultural health, food safety and quality control projects

Country	Fiscal Year	Project Name	Loan Amount (US$ million)	Safety Component
China	1996	Seed Sector Commercialization	100	Lab, external quality control and variety evaluation, seed certification
China	1998	Sustainable Coastal Resource Development	100	Seafood product quality and safety
Cote Divoire	1995	Agricultural Export Promotion and Diversification	5.8	Product development and quality control via adaptive research
Guinea	1993	National Agricultural Export Promotion	20.8	Quality improvement, quality labels to products that are free of chemicals
Hungary	1988	Agroprocessing Modernization	70	Grading, standardizing, quality control, testing products for residues.
Hungary	1990	Integrated Agricultural Export	100	Quality products for export
Kyrgyz Republic	1996	Sheep Development	11.6	Production of quality lambs and wool for premium prices
Madagascar	1993	Food Security and Nutrition	21.3	Campaign to eradicate iodine deficiency disorders
Mali	1995	Agricultural Trading and Processing Promotion Pilot	6	Ensure more uniform product quality and high standards
Moldova	1996	First Agriculture	10	Vinestock certification, product quality and integrated pest management
Morocco	1999	Pilot Fisheries Development	5.2	Quality assurance on-board vessels, spread of good hygienic practice
Myanmar	1985	Second Seed Development	14.5	Seed quality control
Nigeria	1990	National Seed and Quarantine	14	Seed legislation, effective certification and quality control
Peru	1994	Basic Health and Nutrition	34	Micronutrient supplementation and food quality control
Philippines	1990	Small Coconut Farms Development	121.8	Quality management, aflatoxin laboratories, survey of aflatoxin origins
Rwanda	1989	Agricultural Services II	19.9	Improve seed quality and certification
Senegal	1998	Agricultural Export Promotion	8	Quality control; international iso certifications for product processing

Table 3 Agricultural health, food safety and quality control projects

Country	Fiscal Year	Project Name	Loan Amount (US$ million)	Safety Component
Uganda	1995	District Health Services Pilot and Demonstration Project	45	Fortification
Uzbekistan	1995	Cotton Sub-Sector Improvement	66	Grading, seed certification and seed quality control

Table 4 Components of agricultural safety projects

Project Category	*Examples of Project Components*
Food Safety	• Public laboratory equipment for testing pesticide residues, microbial counts, aflatoxin, food adulteration • Marketing infrastructure investments to permit meeting sanitary standards in export markets, e.g. cold storage • Training for producers, e.g. HACCP training for fishery exporters, hygiene for street food vendors • Training for public inspectors and laboratory technicians • Legal framework and enforcement • Phased implementation of HACCP regulation for food processors
Quality Control	• Investment in physical infrastructure for marketing and processing, e.g. packing methods, cold storage, ice making machines for fishing boats • Technical assistance in setting up export marketing • Market information about export market opportunities, incl. standards and other NTB requirements, often through establishing a public export promotion agency • Public laboratory equipment for testing for pesticide residues, microbial counts • Setting up grading system and standards for raw materials • Facilitating establishment of producer groups to provide quality control and certification • Research to support quality improvement
Animal Health	• Public laboratories for animal disease surveillance • Vaccination campaigns • Privatization of veterinary services and/or cost recovery for services such as vaccinations, AI, laboratory tests • Disease eradication campaigns, incl. FMD, tsetse fly, brucellosis • Strengthening laws for quarantine and inspection • Improved extension through training and participatory approaches • Deregulation of veterinary pharmaceutical imports or pricing • Training of veterinary service providers or extension personnel • Promotion of livestock exports a frequent goal • Strengthen legal and institutional capacity for border sanitary control, incl. quarantine and inspection
Plant Health	• Provision of disease free stock for perennials (e.g. fruit, vinestock) • Eradication or control of plant diseases or pests (e.g. Mediterranean fruit fly, locusts) • Public laboratories for disease surveillance, pesticide residues • Capacity building for quarantine and control at borders • Capacity building for seed inspection, certification, labeling • Legal framework for seed inspection, quality control, plant quarantine • Participation in international organizations (UPOV and ISTA) • Integrated Pest Management for plant disease or pest control

Source: World Bank project documents.

7. Potential Expertise and Partnerships Available

International, regional, and subregional organizations serve several vital functions in the areas of agricultural health and food safety, including: establishing trade rules, dispute settlement, standard setting, advisory services, technical assistance, training, research, surveillance, and coordination with international, regional, and national entities. The Bank has begun to engage in discussions with several of these entities regarding their food safety activities. As the Bank explores its involvement in the food safety arena, it will benefit from working cooperatively with international organizations and other partners to strengthen global and regional food safety efforts and enhance its effectiveness in supporting the needs of its clients. A summary of organization and their areas of involvement is provided in table 5 at the end of this chapter.

International Organizations

The **World Trade Organization (WTO),**[3] the multilateral institution charged with administering the WTO Agreements, serves as a forum for trade negotiations, monitors country trade policies, provides technical assistance and training for developing countries and cooperates with other international organizations. The SPS Agreement established rules that set a scientific standard for measures whose purpose is to protect: human or animal health from foodborne risks; human health from animal- or plant-carried diseases; or animals and plants from pests or diseases; whether or not these are technical requirements. This agreement includes standards on chemicals, pesticides, food additives, and veterinary drugs. The related WTO Agreement on Technical Barriers to Trade (TBT Agreement) applies to all aspects of food standards not covered by the SPS Agreement. These include labeling requirements, nutrition claims, quality and packaging regulations, which are generally not considered to be sanitary or phytosanitary measures.

The SPS Agreement supports international harmonization of national measures on the standards, guidelines and recommendations developed by WTO member governments in other international organizations. These organizations include: the Joint FAO/WHO CODEX Alimentarius Commission (CODEX) for food safety, Office International des Epizooties (OIE) for animal health, and International Plant Protection Convention (IPPC) for plant health. These organizations are expected to aid in the settlement of agricultural health and food safety disputes through the promotion of a rule-based system.

---

[3] WTO website: http://www.wto.org/wto/index.htm viewed 5/17/99.

WTO technical cooperation and training activities aim to assist countries in their understanding and implementation of agreed international trade rules, in participating more effectively in the multilateral trading system and ensuring a lasting, structural impact by contributing to human resource development and institutional capacity building. Six core organizations are working together with the Least Developed Countries to coordinate their trade assistance programs through an Integrated Framework for Trade-Related Technical Assistance to these countries. These agencies are the International Monetary Fund (IMF), the International Trade Centre (ITC), the United Nations Conference on Trade and Development (UNCTAD), the United Nations Development Program (UNDP), the World Bank, and the World Trade Organization (WTO). The program has two components:

- creating autonomous commitments by WTO member countries to provide enhanced market access for least developed country exports, and;

- organizing a program among the international financial institutions, including the World Bank, and aid agencies to provide trade-related assistance.

The WTO and the World Bank Institute (WBI) jointly run the Trade and Development Centre, which provides internet resources on trade and developing countries, including case studies, information about Codex Alimentarius, on-line discussion forums on trade and development, and interactive guides on WTO issues. Activities also include WTO training activities, which aim to assist developing countries in their understanding and implementation of WTO trade rules. The WTO finances and organizes Trade Policy Courses for trade officials from developing countries and economies in transition which are in the process of accession to the WTO. One initiative—*Wiring up Africa: Information Technologies for Development (ITD)*—aims to integrate the least-developed and developing country members of the organization into the multilateral trading system.

The Food and Agriculture Organization of the United Nations (FAO)

FAO[4] provides advice and technical assistance to its members on: analytical methodologies and agricultural policy analysis concerning trade issues, development of food quality control strategies, strengthening national food control systems, reformulating national food regulations to bring them into conformity with international standards, establishing import/export food inspection and certification programs, education and training in agricultural safety development of programs to assist governments, food producers, manufacturers and processors.

FAO develops and implements programs to control environmental contaminants in food.. FAO works closely with the World Health Organization (WHO), United Nations Environment Program, WTO, the International Agency for Research on Cancer (IARC) and the International Atomic Energy Agency (IAEA) on technical matters relating to agricultural safety, quality and consumer protection.

---

[4] FAO website: http://www.fao.org/ viewed 5/21/99.

FAO's Food Quality and Standards Service assists developing countries in establishing or strengthening national food control systems. They implement projects to improve laws and regulations, inspection, certification, monitoring and training. FAO produces technical materials on food control. They sponsor events such as recent expert consultations on food quality and safety, which focused on biotechnology, risk analysis, risk management and risk communication.

The FAO Plant Protection technical assistance program addresses: capacity building; increased understanding of trade related principles of plant protection; and development and upgrade of the legislative framework for plant quarantine. The Animal Health Service (AGAH) is responsible for helping member countries to develop strategies for the economic control of animal diseases.

Agro-Industries and Post-Harvest Management Service (AGSI) carries out activities in the food processing sector aimed at upgrading food processing/preservation technologies, improving food quality, and adding value to raw agricultural materials. AGSI also works on quality assurance, packaging, new and emerging technologies and biotechnologies applicable to food preservation.

The **Joint FAO/WHO CODEX Alimentarius Commission (CODEX)**[5] is recognized in the SPS agreement as the source of standards, guidelines, and recommendations regarding food safety standards. CODEX comprises 162 member countries and has formulated many standards for a wide range of food commodities, food safety, pesticide residues, food additives, veterinary drug residues, food contaminants and labeling. It has also elaborated Codes of Hygienic Practice and Principles for food import and export inspection and certification. Codex performs its work through committees composed of delegates from member countries. There are general subject committees and commodity committees. CODEX organized Regional Workshops on the Establishment and Administration of a National Codex Committee for the English-speaking Caribbean and Central and Eastern European Countries and members of the Commonwealth of Independent States to facilitate countries' participation in Codex work, and to promote the utilization of Codex work at the national level and in international food trade.

The **International Plant Protection Convention (IPPC)**[6] is a multilateral treaty whose purpose is to prevent the spread and introduction of plant pests and to promote measures for their control. It provides a framework and forum for international cooperation, harmonization and technical exchange in collaboration with regional and national plant protection organizations. The IPPC is recognized by the WTO in the SPS Agreement as the source for international standards for the phytosanitary measures affecting trade. The IPPC Secretariat provides technical advice and assistance to governments with the aim of improving the effectiveness of their national plant projection organizations and their capacity in modern technologies, such as pest risk analysis and in international trade.

---

[5] CODEX website: http://www.fao.org/WAICENT/faoinfo/economic/ESN/codex/Default.htm viewed 5/17/99.

[6] IPPC website: http://www.fao.org/waicent/faoinfo/agricult/agp/agpp/pq/default.htm viewed 5/17/99

These assistance activities are coordinated through the FAO Plant Protection Service (AGPP).

Regional Plant Protection Organizations (RPPOs)

The RPPOs function as coordinating bodies in the different continents to further the objectives of the International Plant Protection Convention, and to gather and disseminate information. These include:

- *NAPPO: North American Regional Plant Protection*

- *COSAVE: Regional Plant Health Committee of the Southern Cone*

- *OIRSA: Regional Organization for Agricultural Health (Mexico, Guatemala, El Salvador, Honduras, Nicaragua, Costa Rica y Panama, Dominican Republic)*

- *APPPC: Asia and Pacific Plant Protection Commission*

- *CPPC: Caribbean Plant Protection Commission*

- *CA: Andean Community*

- *EPPO: European and Mediterranean Plant Protection Organization*

- *IAPSC: Interafrican Phytosanitary Council*

- *PPPO: Pacific Plant Protection Organization*

Each RPPO has its own independent statutes and conducts its own regional cooperation program. RPPOs produce regional standards for their members. They cooperate with each other and with FAO. In particular, they meet in Technical Consultations to promote the development and use of relevant standards and to encourage inter-regional cooperation on phytosanitary measures for controlling quarantine pests and preventing their introduction and spread.

The **Office International des Epizooties (OIE)**[7] is also known as the World Animal Health Organization and has the objectives of: informing Governments of the occurrence of animal diseases throughout the world, and of ways to control these diseases; coordinating international studies for surveillance and control of animal diseases; and harmonizing regulations for trade in animals and animal products among the 151 Member Countries. The OIE International Committee promotes the harmonization of regulations applicable to trade in animals and animal products through the following:

- The *International Animal Health Code for Mammals, Birds and Bees*, which provides standards for international trade;

---

[7] OIE website: http://www.oie.int/A_html.htm viewed 5/17/99.

- The *Manual of Standards for Diagnostic Tests and Vaccines*, which provides the standardized diagnostic techniques and vaccine control methods for use in international trade;

- The *International Aquatic Animal Health Code and Diagnostic Manual for Aquatic Animal Diseases*, which contains general information on international trade in fish, molluscs and crustaceans, including sections on import risk analysis and import/export procedures.

The **International Organization for Standardization (ISO)**[8] is a worldwide federation of national standards bodies from 130 countries, which seeks to promote the development of standardization and related activities to facilitate international trade, and to develop cooperation in intellectual, scientific, technological and economic activity. ISO is recognized as providing a special technical support role in relationship to WTO. The TBT Agreement encourages Members to participate in the work of international bodies for the preparation of standards and guides or recommendations for conformity assessment procedures. ISO offers a wide array of standardized sampling, testing and analytical methods and is developing environmental management system standards that can be implemented in any type of organization. ISO coordinates its activities with CODEX and invites CODEX to sit in on the deliberations of Technical Committee 34, which handles technical standards for agricultural food products.

The ISO 9000 family of standards represents an international consensus on good management practices with the aim of ensuring that the organization can consistently deliver the product or services that meet the client's quality requirements. ISO 14000 is primarily concerned with how the organization manages its activities to minimize harmful effects on the environment.

ISO conducts training seminars, publishes development manuals, and provides various other kinds of expert assistance for its developing country members. Its work, which is supported by governmental aid agencies and ISO members from several industrialized countries, assists developing countries in the development of their national standardization and quality assurance systems.

The **World Health Organization (WHO)**[9] is involved in the development of national agricultural safety policies and infrastructures, reviewing and assessing local needs, and intersectoral collaboration for implementing agricultural safety activities. They deal with food legislation and enforcement, promoting awareness of food processing technologies that will help prevent foodborne disease and decrease postharvest losses, agricultural safety education, epidemiological surveillance of foodborne diseases and monitoring agricultural safety infrastructure. WHO implements the Global Environment Monitoring System—Food Contamination Monitoring and Assessment Program (GEMS/Food), which informs governments, the Codex Alimentarius Commission, other relevant

---

[8] ISO Website: http://www.iso.ch/infoe/intro.htm#What is ISO, viewed 5/11/99.

[9] WHO website: http://www.who.int/fsf/fctshtfs.htm#Role of WHO in promotion of food safety, viewed 5/21/99.

institutions, and the public, on levels of contaminants in food and significance with regard to public health and trade. WHO's food safety work is coordinated at Headquarters by the Food Safety Programme, Department of Protection of the Human Environment, Cluster on Sustainable Development and Healthy Environments (FOS/PHE/SDE) and, at the regional and country level, by Regional Advisers. WHO collaborates with the Industry Council for Development (ICD) and the International Life Sciences Institute (ILSI), both NGOs. Examples of joint activities include an Asian Conference on Food safety, training in HACCP in several countries, training of nutritionists in food safety in Indonesia and development of a food safety program in Indonesia.

The **International Trade Centre UNCTAD/WTO (ITC)**[10] coordinates technical cooperation with developing countries in trade promotion. ITC works closely with UNCTAD and the WTO on its technical cooperation programs. ITC's export market development activities are coordinated with the work of the FAO and the United Nations Industrial Development Organization (UNIDO). ITC, UNCTAD and WTO have launched a joint training program for trainers in developing countries to give a comprehensive overview to Multilateral Trading System issues. ITC has held national and regional workshops on the TBT and SPS agreements and the ISO 9000 and 14000 series of standards and published handbooks on export quality issues.

The **International Atomic Energy Agency (IAEA)**[11] undertakes a variety of technical assistance and cooperation activities relating to animal and plant health and food safety and quality. Activities include: eradication of insect pests through sterile insect technique; animal disease surveillance, which includes Rinderpest surveillance and expert services, including a full time regional expert to ensure the successful transfer of the ELISA (enzyme linked immunosorbent assay) technique and the routine use of the ELISA based system; equipment where necessary; and annual workshops for dissemination of national seromonitoring results on a regional basis; and nutrition and human health activities in three different areas: human nutrition, environmental studies of non-radioactive pollutants (studied by nuclear techniques) and monitoring of accidentally released radionuclides in environmental and food samples. All of these activities are supported through coordinated research, technical cooperation, and other means. Some of them are also supported through activities in the Agency's Laboratories.

Regional Organizations

Various regional organizations have tried to coordinate their aid efforts in providing institutional support so that their regional member countries might better be able to meet health and safety concerns and WTO standards.

---

[10] International Trade Centre website: http://www.intracen.org/welcome.htm viewed 5/7/99.

[11] IAEA website: http://www.iaea.org/worldatom/inforesource/other/tc_90s/chpthree.html viewed 5/19/99.

The Inter-American Institute for Cooperation on Agriculture (IICA)[12] has been a leader in food safety and agricultural health for its 34 member states in the Americas and is working to build global partnerships with international organizations, national agencies, research and educational institutions, and private sector entities. It cooperates with national agricultural health systems in modernizing their infrastructure, organization and operations so that they can ensure that countries develop and maintain optimal agricultural health and food safety conditions and comply with regional and international commitments.

IICA and the Mexican Department of Agriculture (SAGAR), sponsored an international conference on Food Safety in September, 1998, which was attended by about 350 people from 20 countries. They hosted a follow-up workshop, which involved experts from the Americas in food safety and animal and plant health. The objective of that workshop was to design a basic model that would serve as a guide for countries to modernize their agricultural health and food safety systems to comply with and take advantage of food safety, animal and plant health requirements. IICA now hopes to foster partnerships among critical organizations such as the World Bank, IDB, PAHO, and WTO, among others. IICA cooperated with the World Bank and EMBRAPA in organizing a Hemispheric Conference titled: *Future Food Safety Strategies: Collaborative Roles Between International Agencies, Public and Private Sectors.*[13]

The **Asia-Pacific Economic Cooperation (APEC)**[14] has become a key regional vehicle for promoting open trade and economic cooperation. The Agricultural Technical Cooperation Experts Group (ATC), focuses on seven priority areas: conservation and utilization of plant and animal genetic resources; research, development and extension of agricultural biotechnology; marketing, processing and distribution of agricultural products; plant and animal quarantine and pest management; cooperative development of an agricultural finance system; agricultural technology transfer and training; and sustainable agriculture. They organize workshops on risk assessment of plant pests, among other activities. A fisheries working group conducted several technical workshops on seafood health and quality rules and technical workshops on the principles of HACCP- based seafood inspection.

Pan American Health Organization (PAHO)

The PAHO[15]/WHO Veterinary Public Health Program aims to support the national priority programs of the member governments in:

- the promotion of animal health

---

[12] IICA website: http://www.iica.ac.cr/ viewed 5/28/99.

[13] The conference report can be viewed at the IICA web site: http://www.iica.ac.cr/sanidad.

[14] APEC website: http://www1.apecsec.org.sg/97brochure/97brochure.html viewed 4/14/99.

[15] PAHO website: http://www.paho.org/ viewed 5/17/8/99.

- the protection of food for human consumption to guarantee its safety and nutritional quality and prevent the transmission of disease agents through this medium;

- the surveillance, prevention, and control of zoonoses and communicable disease common to humans and animals that cause widespread morbidity, disability, and mortality in vulnerable human populations;

- the promotion of environmental protection in regard to the potential risks to public health.

PAHO convenes ministerial meetings for Health and Agriculture, in order to foster integrated national programs for food protection (PAHO). They provide technical cooperation in the organization of integrated national programs, strengthening analytical capacity in monitoring, strengthening of inspection services, and in establishing epidemiological surveillance of foodborne diseases.

The **Association of Southeast Asian Nations (ASEAN)**[16] has as one of its strategic thrusts the enhancement of intra- and extra-ASEAN trade and long-term competitiveness of ASEAN's food and agricultural products/commodities. Their work includes activities to intensify cooperation in production and processing technology development and enhancement of development, harmonization and adoption of quality standards for products through:

- Development and adoption of quality assurance systems for selected tropical fruits;

- Implementation of a common protocol for irradiation as a quarantine treatment for the trade of fresh fruits and vegetables and adoption of a harmonized regulation on food irradiation for ASEAN;

- Harmonization of phytosanitary measures for crop products;

- Harmonization of maximum residue limits (MRLs) of commonly used pesticides for vegetables traded among ASEAN Member Countries;

- Establishment of an accreditation scheme for establishments involved in the production of livestock products traded among ASEAN Member Countries;

- Standardization of quality control measures and processing techniques for fish and fisheries products;

- Harmonization of fisheries sanitary measures among ASEAN Member Countries; and

---

[16] ASEAN website: http://www.aseansec.org/ viewed 5/28/99.

- Harmonization of regulations for agricultural products derived from biotechnology, among other activities.

Private Sector Organizations

The private sector plays an important function in the development and implementation of animal and plant health and food safety measures. Many private sector organizations participate actively in the work of Codex and other standard setting organizations. In addition, many organizations carry out training and technical assistance activities to assist private sector organizations to implement food safety measures.

It would be impossible to catalogue all of the private sector activities in the food safety area. The following are a few examples of active organizations that may be potential partners in World Bank activities.

International

The International Life Sciences Institute (ILSI) is a worldwide foundation that aims to enhance public health and advance knowledge related to nutrition, food safety, toxicology and the environment by supporting long-term scientific research in underfunded or unexplored areas. ILSI has NGO status with WHO and specialized consultative status with FAO. ILSI collaborates with leading international health and development organizations in projects that encourage global cooperation to develop harmonized or equivalent science-based standards. ILSI's branches sponsor scientific research, conferences, workshops, symposia, and publications.[17] Recent food safety activities include: a workshop in Brazil to review scientific issues related to the safety assessment of foods derived from Genetically Modified Plants; a March 1999 Conference on Emerging Foodborne Pathogens: Implications and Control, which was ILSI's largest United States-based conference to date. ILSI has several technical committees working on issues related to food safety, including: Antioxidant Technical Committee, Technical Committee on Food Microbiology, and the Technical Committee on Food Toxicology and Safety Assessment.

The **International Food Information Service (IFIS)** provides information on food science, food technology, and human nutrition for those working in the food sector.[18] They maintain databases of food science alerts, *Food Science and Technology Abstracts*, a food and nutrition internet index and other information resources.

Industry Council For Development (ICD)[19] collaborates with WHO as well as bilateral and private development organizations to assist developing countries achieve their social

---

[17] ILSI Website: http://www.ilsi.org viewed 4/14/99.

[18] IFIS Website: http://www.ifis.org viewed 4/14/99.

[19] WHO Website: http://www.who.int/ina-ngo/ngo/*ngo015*.htm viewed 5/21/99.

and economic goals. They contribute industry's managerial, technological, scientific and other resources in accordance with national development objectives.

The International Association of Milk, Food and Environmental Sanitarians (IAMFES) [20] is a nonprofit association of food safety professionals. Comprised of over 2,900 Members from 50 nations, IAMFES is recognized as having the leading food safety conference, and is dedicated to advancing food safety worldwide. IAMFES publishes the *Journal of Food Protection*, a refereed monthly publication of scientific research and authoritative review containing articles on a variety of topics in food science related to food safety and quality.

Consumers International[21] is a worldwide nonprofit federation of consumer organizations, dedicated to the protection and promotion of consumer interests. It is involved in monitoring the establishment of food standards at the international level through the CODEX. It also has official representation on WHO and ISO. It works to ensure that consumers' interests on food irradiation and biotechnology issues are recognized. Consumers International has co-founded three global health networks: the International Baby Food Action Network; the Health Action International (campaigning for fairer health and drug policies) and The Pesticides Action Network. It issues a broad range of publications on food safety issues.

National

United States

The **Food Marketing Institute (FMI)** [22] is a nonprofit association conducting programs in research, education, industry relations and public affairs on behalf of its members - food retailers and wholesalers and their customers in the United States and around the world. FMI has undertaking technical assistance programs in food safety in cooperation with the government and private sector entities in developing countries. They have been active in developing a strategy to minimize foodborne illness in fresh raspberries in Guatemala.

The International Food Information Council (IFIC)[23] is a nonprofit foundation based in Washington, D.C. that provides scientific information on agricultural safety and nutrition to journalists, health professionals, educators, government officials and consumers.

---

[20] IAMFES Website: http://www.iamfes.org/ viewed 4/15/99.

[21] Consumers International Website http://193.128.6.150/consumers//index.html viewed 4/15/99.

[22] Food Marketing Institute Website: http://www.fmi.org viewed 4/14/99.

[23] IFIC Website: http://www.ificinfo.health.org viewed 4/14/99.

National Food Processors' Association (NFPA)[24] is the principal scientific and technical trade association for the food industry and provides support to its members on agricultural safety, processing, health and nutrition issues. NFPA's Office of Food Safety Programs focuses on integrating the Association's resources in microbiology, risk assessment, and other technical areas related to food safety, including the use of HACCP programs by the food industry. Its education provider, the Food Processors Institute (FPI), publishes its *HACCP Manual* and offers company-specific and industry-wide training in HACCP.

Grocery Manufacturers' Association (GMA)[25] is active in the area of science and the regulatory process, and the interpretation of multilateral trade agreements. Its activities include: 1) Participation in Codex policy development; direction on the reorganization of the U.S. Codex Office; leadership to the Food Industry Codex Coalition, 2) Advocacy on important industry trade issues before the United States Trade Representative (USTR) and the WTO, 3) Monitoring and responding to major legal/regulatory changes in other important trading countries via the U.S. Department of Agriculture's Foreign Agricultural Service, and 4) Building communications with other countries' food trade associations to help bring alignment worldwide in support of science-based Codex decisions.

NSF International[26] is an independent, nonprofit organization involved in the development of standards, product testing and certification in the areas of public health safety and protection of the environment. NSF's principal services include management systems assessment and registration (ISO 9000 standards for quality, ISO 14000 standards for environment, and HACCP-9000® for food safety and quality); and training in HACCP and Food Safety.

Australia

Australia Food Safety Campaign Group[27] is a partnership among government, industry, consumer and professional associations that aims to educate the Australian public on food safety. It carries out a mass communication program to raise the awareness among consumers and food handlers of the need for safe food handling practices.

New Zealand

Meat New Zealand[28] is the New Zealand meat industry organization that does a variety of activities relating to food safety, including:

---

[24] NFPA Website http://www.nfpa-food.org viewed 4/14/99.

[25] GMA Website: http://www.gmabrands.com viewed 4/14/99.

[26] NSF International Website: http://www.nsf.org/index.html viewed 4/20/99.

[27] Australia Food Safety Campaign Website: http://www.safefood.net.au/index.cfm viewed 4/14/99.

[28] Meat New Zealand Website: http://www.nzmeat.co.nz/ viewed 5/19/99.

- on-going review of food safety systems for known food safety risks

- developing scenarios for responding to potential food safety failures in key markets

- developing and implementing product traceability and quality management systems

- encouraging the use of HACCP-type safety systems

- developing technologies that ensure the microbial, chemical and physical safety of meat; and

- providing consumers with food handling guidelines.

Canada

Canadian Partnership for Consumer Food Safety Education[29] is a nonprofit organization of committed to reducing foodborne illness in Canada by increasing awareness of safe food handling practices through the coordination and delivery of food safety education programs focused on the consumer. Membership includes over 20 industry, consumer, health and environmental organizations, and the federal and provincial government organizations who are concerned with food safety.

United Kingdom

The **Royal Institute of Public Health & Hygiene and the Society of Public Health (RIPHH)**[30] is a nonprofit organization whose mission is to advance the cause of health and hygiene. They provide food hygiene, safety and HACCP training for food handlers, supervisors, and managers. RIPHH food hygiene and safety standards are designed to meet the needs of all types of food business. RIPHH overseas examinations are held regularly in Hungary, Turkey, Cyprus, the Gulf, India, Canada and Ireland.

Chile

Fundación Chile[31] is a nonprofit organization devoted to technological innovation through the transfer of technologies that contribute to enhanced use of natural resources and increased productive capacity. They provide technical assistance services, training, consulting, and technology transfer through creation, development and sale of innovative businesses. They conduct an international course on food canning for companies

---

[29] Canadian Partnership for Consumer Food Safety Education website: http://www.canfightbac.org/english/ viewed 4/15/99.

[30] The Royal Institute of Public Health & Hygiene and the Society of Public Health website: http://www.riphh.org.uk/index.html viewed 4/20/99.

[31] Fundación Chile Website: http://www.fundch.cl/fc/index.htm viewed 5/19/99.

interested in exporting to the United States. They also carry out quality assurance, food safety and training for produce and aquaculture products. The quality control laboratory sells inspection and certification services to the Chilean food industry for domestic and export markets.

Bilateral Partners

The governments of the United States, European Union, Canada, and other countries play major roles in the areas of international plant health, animal health, and food safety and would be able to contribute significant support to Bank activities. Country agencies involved in plant health are responsible for developing the legal framework for plant protection and the development of phytosanitary standards. Part of their role is to monitor fruit and vegetable products for pest and disease by providing identification, surveillance, quarantine, and control and eradication systems when needed. Those agencies involved in animal health are responsible for the developing the legal framework and zoosanitary standards for the country. They are also responsible for maintaining diagnostic, quarantine, and information systems on outbreaks. They develop survey techniques, control methods, and training on epidemiology and disease surveillance. Those agencies involved in food safety provide technical assistance to producers on the safe handling of foods throughout the distribution chain, capacity building, and training on inspection systems. A variety of country agencies also work on liaison and capacity building with international and regional organizations and provide consumer education programs.

Table 5 Organizations involved in animal and plant health, food safety, and quality control

Organization	Food safety	Animal health	Plant health	Quality control	Support to private production/ processing	Standard setting, harmonization, and/or dispute settlement	Institutional or regulatory capacity building
APEC	X	X	X	X	X	X	X
ASEAN	X	X	X	X	X	X	X
CODEX	X					X	X
FAO	X	X	X	X	X		
IAEA	X	X	X	X	X		X
IAMFES	X			X	X		
ICD	X						X
IFIS	X			X	X		
IICA	X	X	X		X		X
ILSI	X				X		X
IPPC			X			X	X
ISO				X	X	X	X
ITC	X			X	X		X
OIE		X				X	X
PAHO	X	X					X
WHO	X			X			X
WTO	X	X	X	X		X	X

Source: See text discussion for institutional sources.

References

American Chemical Society. *Understanding Risk Analysis*. 1998. Available on the web
 http://www.rff.org/misc_docs/risk_book.htm

Anderson, C. 1991. "Cholera Epidemic Traced to Risk Miscalculation", *Nature*, Vol.
 354, 28 November 1991, p. 255.

Ghat, R.V. and S. Vasanthi. "Mycotoxin Contamination of Foods and Feeds", Third Joint
 FAO/WHO/UNEP International Conference on Mycotoxins, Tunisia, 3-6 March,
 1999. Available on the web at
 http://www.fao.org/waicent/faoinfo/economic/ESN/mycoto/papers/myco4a.pdf

Binswanger, H.P. and P. Landell-Mills. 1995. *The World Bank's Strategy for Reducing
 Poverty and Hunger*. Environmentally Sustainable Development Studies and
 Monographs Series No. 4. Washington, D.C.: The World Bank.

Binswanger, H.P. and J.Von Braun. 1991. "Technological Change and
 Commercialization in Agriculture: The Effect on the Poor". *World Bank
 Research Observer*. January 1991, Vol. 6: 57-80.

Bureau, J. and S. Marette. 1999. "Accounting for Consumers Preferences in
 International Trade", Paper presented at the National Research Council
 Conference on Incorporating Science, Economics, Sociology and Politics in
 Sanitary and Phytosanitary Standards in International Trade, Irvine, CA, Jan 25-
 27, 1999.

Buzby, J.C., Roberts, T., Lin, C.-T.J., and MacDonald, J.M. 1996. Bacterial
 Foodborne Disease: Medical Costs and Productivity Losses. Economic Research
 Service, U. S. Department of Agriculture, *Agricultural Economic Report
 No. 741*.

Council for Agricultural Science and Technology (CAST). 1994. *Foodborne
 Pathogens: Risks and Consequences*. R122. September 1994.

Caswell, J.A., ed. 1995. *Valuing Food Safety and Nutrition*. Westview Press.

Cato, J.C. and Dos Santos, C.A.L. 1999. "Costs to Upgrade the Bangladesh Frozen
 Shrimp Processing Sector to Adequate Technical and Sanitary Standards" in *The
 Economics of HACCP*, ed. Laurian Unnevehr, St. Paul: Eagan Press.
 forthcoming.

Coirolo, Luis O. 1999. "Brazil Animal and Plant Protection Project", Presentation World Bank Rural Week, March 24-26, 1999.

Cropper, M.L.. 1995. "Valuing Food Safety: Which Approaches to Use?", in Caswell, J.A., ed. *Valuing Food Safety and Nutrition.* 1995. Westview Press.

Food and Agriculture Organization of the United Nations. 1999. "The Importance of Food Quality and Safety for Developing Countries". Committee on World Food Security, 25th Session, Rome, May 31-June3, 1999, at http://www.fao.org/docrep/meeting/x1845e.htm

Hammer, J.S. 1997. "Economic Analysis for Health Projects". *The World Bank Research Observer.* Vol. 12, No. 1 (February 1997), pp 47-71.

Hathaway, Steve C. 1995. Harmonization of International Requirements Under HACCP-Based Food Control Systems. *Food Control* 6. pp. 267-76.

Hathaway, S.C. and R.L. Cook. 1997. "A Regulatory Perspective On The Potential Uses Of Microbial Risk Assessment In International Trade". *International Journal of Food Microbiology* 36 (1997) pp. 127-133.

Institute for Inter-American Cooperation in Agriculture (IICA). 1999. "Future Food Safety Strategies: Collaborative Roles Between International Agencies, Public and Private Sectors". Report from an IICA/EMBRAPA/IBRD Conference, San Jose, Costa Rica, August 26-27, 1999. Available on the web at http://www.iica.ac.cr/sanidad.

International Commission on Microbiological Specifications for Foods (ICMSF). 1988. *Application of the Hazard Analysis Critical Control Point (HACCP) System to Ensure Microbiological Safety and Quality.* **Microorganisms in Food 4.** Blackwell Scientific Publications, Oxford.

International Commission on Microbiological Specifications for Foods (ICMSF) (1997) Establishment of microbiological safety criteria for foods in international trade. *World Health Statist.* **50.** 119-23.

International Commission on Microbiological Specifications for Foods (ICMSF). 1998. "Potential Application of Risk Assessment Techniques to Microbiological Issues Related to International Trade in Food and Food Products". *Journal of Food Protection* 61. pp. 1075-1086.

Jaffee, Steven. *Exporting High-Value Food Commodities: Success Stories from Developing Countries.* World Bank Discussion Papers 198.

Jasanoff, S. "Technological Risk and Cultures of Rationality" Paper presented at the National Research Council Conference on Incorporating Science, Economics,

Sociology and Politics in Sanitary and Phytosanitary Standards in International Trade, Irvine, CA, Jan 25-27, 1999.

Kaferstein, F. and M. Abdussalam. 1998. "Food Safety in the Twenty-first Century", Proceedings of the 4th World Congress on Foodborne Infections and Intoxications". Berlin Germany, June 1998.

Lvovsky, K., G. Hughes, and M. Dunleavy. 1999. *Investing in People Through Investing in Water*. Discussion Paper World Bank, March 1999.

MacDonald, J.M. and S. Crutchfield. 1996. "Modeling the Costs of Food Safety Regulation." *American Journal of Agricultural Economics*. Vol. 78, December 1996, pp. 1285-90.

Mazzocco, Michael. 1996. HACCP as a Business Management Tool. *American Journal of Agricultural Economics* 78. 770-774.

Meerman, J. 1997. *Reforming Agriculture: The World Bank Goes to Market*. The World Bank, Washington DC.

Motarjemi,Y. and F.K. Kaferstein. 1997. "Global Estimation of Foodborne Diseases. *World Health Statistics Quarterly*, 50(1997): 5-11.

Motarjemi, Y., F. Kaferstein, G. Moy, and F. Quevedo. 1993. "Contaminated Weaning Food: a Major Risk Factor for Diarrhoea and Associated Malnutrition". *Bulletin of the World Health Organization*. 71 (1): 79-92.

Motarjemi, Y., Käferstein,F., Moy, G., Miyagawa, S., and Miyagishima, K. 1996. "Importance of HACCP for Public Health and Development: the Role of the World Health Organization". *Food Control* 7; pp. 77-85.

Moy, G., Hazzard, A., and Kaferstein, F. 1997. "Improving the Safety of Street-Vended Food". *World Health Statistics Quarterly* 50; pp. 124-131.

National Advisory Committee on Microbiological Criteria for Foods (NACMCF). 1992. "Hazard Analysis and Critical Control Point System". *International Journal of Food Microbiology* 16; pp. 1-23.

PAHO/WHO. 1997. *Integration of Food Protection in the Region of the Americas*. Report from the X Inter-American Meeting at the Ministerial Level on Animal Health, Washington DC, 23-25 April, 1997.

Pierson, M.D. and Corlett, Jr., D.A., eds. 1992. *HACCP: Principles and Applications*. New York: Van Nostrand Reinhold.

Roberts, D. 1998. "Preliminary Assessment of the Effects of the WTO Agreement on Sanitary and Phytosanitary Trade Regulations". *Journal of International Economic Law*. pp. 377-405.

Roberts, T., K.D. Murrell, and S. Marks. 1994. "Economic Losses Caused by Foodborne Parasitic Diseases". *Parasitology Today*, Vol. 10, No. 11, pp. 419-423.

Simeon, Michel. 1999. "Food Safety and Quality Management". Presentation World Bank Rural Week, March 24-26, 1999.

Sullivan, G.H., G.E. Sanchez, S.C. Weller, and C.R. Edwards. 1999. "Sustainable Development in Central America's Non-Traditional Export Crops Sector Through Adoption of Integrated Pest Management Practices: Guatemalan Case Study" *Sustainable Development International*. pp. 123-126.

Tauxe, R.V. 1997. "Emerging Foodborne Diseases: An Evolving Public Health Challenge", *Emerging Infectious Diseases*, Vol. 3, No. 4, Oct-Dec, pp. 425-434.

Thrupp, L. 1995. *Bittersweet Harvests for Global Supermarkets: Challenges in Latin America's Agricultural Export Boom.* World Resources Institute.

Umali, Dina; Gershon Feder, and Cornelis de Haan. 1992. *The Balance Between Public and Private Sector Activities in the Delivery of Livestock Services.* **Discussion Paper 163**, Washington DC: World Bank.

Unnevehr, L.J. and H.H. Jensen. 1996. "HACCP as a Regulatory Innovation to Improve Food Safety in the Meat Industry". *American Journal of Agricultural Economics* 78: 764-769.

Unnevehr, L. J. and H.H. Jensen. 1999. "The Economic Implications of Using HACCP as a Food Safety Regulatory Standard". *Food Policy* 24: 625-635.

U.S. GAO. 1999. *Food Safety: Experiences of Four Countries in Consolidating their Food Safety Systems.* GAO/RCED-99-80, April 1999.

Van Ravenswaay, E.O. and Hoehn, J.P. 1996. The theoretical benefits of food safety policies: A total economic value framework. *American Journal of Agricultural Economics* 78: 1291-1296.

Walker, K. 1999. "Political Dimensions of Food Safety, Trade, and Rural Growth". Presentation World Bank Rural Week, March 24-26, 1999.

World Health Organization. Food Safety—a Worldwide Public Health Issue. Internet Resource. http://www.who.int/fsf/fctshtfs.htm, viewed 3/1/99

Wilson, S. 1999. "China Seafood: Training in Food Quality and Safety". Presentation World Bank Rural Week, March 24-26, 1999.

World Bank. 1993. *World Development Report 1993: Investing in Health.* Washington, D.C.: Oxford University Press for the World Bank.

World Bank. 1995. *Kenya Poverty Assessment.* East Africa Department. Report 13152-KE.

World Bank. 1997. *Rural Development: From Vision to Action.* Environmentally Sustainable Development Studies and Monographs Series 12. Washington DC: The World Bank.

World Bank. *Can the Environment Wait? Priorities for East Asia.* November 1997.

Appendix: Trade Data

Table A1 Fresh food exports from South Asia to industrial economies

(Value in $1000)

Product	Year			
	1993	*1994*	*1995*	*1996*
Meat	504	965	6,222	3,817
Fish	984,095	1,214,089	1,123,923	1,157,923
Fruit	323,382	353,033	274,545	340,258
Vegetable	23,552	34,051	24,802	30,977
Total Fresh	1,331,533	1,602,138	1,429,491	1,532,974
Total Agriculture	2,094,279	2,469,333	2,287,299	2,562,064
Total Exports	20,576,653	23,974,919	24,768,784	26,673,000
Fresh as a percent of agriculture	63.58	64.88	62.50	59.83
Agriculture as a percent of total	10.18	10.30	9.23	9.61
Fresh Exports by Major Market				
EEC-15	481,648	544,313	574,765	568,303
North America	345,137	424,455	330,792	362,774
Japan	508,098	633,836	535,245	604,796

Note: South Asia includes Afghanistan, Bangladesh, Bhutan, India, Maldives, Nepal, Pakistan, and Sri Lanka

Table A2 Fresh food exports from East Asia to industrial economies

(Value in $1000)

Product	Year			
	1993	1994	1995	1996
Meat	1,752,037	2,170,337	2,639,954	2,831,774
Fish	6,180,589	7,115,800	7,592,834	6,667,049
Fruit	571,430	615,255	571,558	612,534
Vegetable	1,605,678	1,589,068	1,527,853	1,544,418
Total Fresh	10,109,734	11,490,461	12,332,199	11,655,776
Total Agriculture	18,494,321	21,440,411	22,031,115	22,267,938
Total Exports	267,918,631	307,889,561	368,543,635	377,585,341
Fresh as a percent of agriculture	54.66	53.59	55.98	52.34
Agriculture as a percent of total	6.90	6.96	5.98	5.90
Fresh Exports by Major Market				
EEC-15	1,517,342	1,470,719	1,464,765	1,462,797
North America	1,457,232	1,616,943	1,485,218	1,412,336
Japan	7,003,648	8,242,600	9,225,633	8,647,537

Note: East Asia includes Cambodia; China; Fiji; Hong Kong, China; Indonesia; Kiribati; Lao People's Democratic Republic; Macau; Malaysia; Papua New Guinea; Philippines; Republic of Korea; Samoa; Singapore; Solomon Islands; Taiwan (China); Thailand; Tonga; Vietnam

Table A3 Fresh food exports from Sub-Saharan Africa to industrial economies

(Value in $1000)

Product	Year			
	1993	**1994**	**1995**	**1996**
Meat	12,549	8,736	15,853	10,037
Fish	112,585	157,141	133,433	156,030
Fruit	21,568	27,274	116,579	111,158
Vegetable	42,073	41,026	75,771	64,499
Total Fresh	188,775	234,176	341,636	341,723
Total Agriculture	1,361,578	1,492,708	2,105,600	1,835,620
Total Exports	6,769,461	6,482,057	6,594,586	7,736,949
Fresh as a percent of agriculture	13.86	15.69	16.23	18.62
Agriculture as a percent of total	20.11	23.03	31.93	23.73
Fresh Exports by Major Market				
EEC-15	176,682	185,692	289,899	296,547
North America	1,563	4,400	15,217	16,847
Japan	17,393	49,583	46,648	39,818

Note: Sub-Saharan Africa includes Angola, Benin, Burkina Faso, Burundi, Cameroon, Cape Verde, Central African Republic, Chad, Comoros, Cote d' Ivoire, Djibouti, Equitorial Guinea, Ethiopia, Gabon, Ghana, Guinea, Guinea-Bissau, Kenya, Liberia, Madagascar, Malawi, Mali, Mauritania, Mauritius, Mozambique, Niger, Nigeria, Republic of Congo, Rwanda, Sao Tome & Principe, Senegal, Seychelles, Sierra Leone, Somalia, Sudan, Tanzania, The Gambia, Togo, Uganda, Zaire, Zambia, and Zimbabwe

Table A4 Fresh food exports from Middle East and North Africa to industrial economies
(Value in $1000)

Product	Year			
	1993	*1994*	*1995*	*1996*
Meat	44,490	45,835	35,782	35,182
Fish	516,072	581,761	710,083	688,893
Fruit	879,260	925,457	1,177,424	1,121,315
Vegetable	429,602	445,244	591,927	594,037
Total Fresh	1,869,424	1,998,296	2,515,215	2,439,427
Total Agriculture	3,243,083	3,573,807	4,210,471	4,158,497
Total Exports	37,081,561	40,463,870	48,855,575	46,828,172
Fresh as a percent of agriculture	57.64	55.92	59.74	58.66
Agriculture as a percent of total	8.75	8.83	8.62	8.88
Fresh Exports by Major Market				
EEC-15	1,528,169	1,595,445	2,007,326	1,965,509
North America	56,089	52,648	73,739	70,083
Japan	264,692	326,421	432,279	390,350

Note: Middle East and North Africa includes Algeria; Arab Republic of Egypt; Islamic Republic of Iran; Iraq; Israel; Jordan; Lebanon; Morocco; Oman; Syria; Tunisia; Turkey; and Republic of Yemen (and People's Democratic Republic Yemen)

Table A5 Fresh food exports from Latin America and the Caribbean to industrial economies

(Value in $1000)

Product	Year			
	1993	1994	1995	1996
Meat	1,142,602	1,307,052	1,352,711	1,362,555
Fish	2,646,678	2,980,390	3,559,856	3,628,820
Fruit	3,574,368	3,695,311	4,093,304	4,363,883
Vegetable	1,515,060	1,641,050	2,120,384	2,028,570
Total Fresh	8,878,707	9,623,803	11,126,256	11,383,829
Total Agriculture	20,208,460	24,041,579	26,785,083	26,941,091
Total Exports	107,719,159	125,000,924	150,283,334	166,821,885
Fresh as a percent of agriculture	43.94	40.03	41.54	42.25
Agriculture as a percent of Total	18.76	19.23	17.82	16.15
Fresh Exports by Major Market				
EEC-15	3,231,957	3,573,598	4,085,944	4,242,138
North America	4,981,718	5,212,549	6,029,344	6,017,713
Japan	646,095	861,490	1,045,083	1,148,604

Note: Latin America and the Caribbean includes Haiti, Antigua and Barbuda, Argentina, The Bahamas, Barbados, Bolivia, Brazil, Chile, Colombia, Costa Rica, Dominica, Ecuador, El Salvador, Grenada, Honduras, Jamaica, Mexico, Nicaragua, Panama, Paraguay, Peru, St. Kitts and Nevis, St. Lucia, St. Vincent and the Grenadines, Suriname, Trinidad and Tobago, Uruguay, Bolivarian Republic of Venezuela, Belize, Guatemala, and Guyana.

Table A6 Fresh food exports from Eastern Europe to industrial economies

(Value in $1000)

Product	Year			
	1993	*1994*	*1995*	*1996*
Meat	540,075	569,742	618,406	635,341
Fish	69,338	96,273	112,090	119,537
Fruit	102,332	118,660	132,447	130,969
Vegetable	278,125	330,547	338,963	307,959
Total Fresh	989,869	1,115,221	1,201,905	1,193,806
Total Agriculture	2,575,199	2,938,851	3,262,535	3,075,971
Total Exports	31,931,586	42,094,977	56,955,308	57,327,600
Fresh as a percent of agriculture	38.44	37.95	36.84	38.81
Agriculture as a percent of total	8.06	6.98	5.73	5.37
Fresh Exports by Major Market				
EEC-15	927,034	1,052,178	1,134,353	1,128,763
North America	10,815	9,339	7,987	6,941
Japan	27,667	24,899	24,954	30,120

Note: Eastern Europe includes Albania; Bulgaria; Boznia and Herzegovina; Czech Republic; Hungary; FYR Macedonia; Poland; Romania; Slovak Republic; Slovenia; and Yugoslavia, Federal Republic of (Serbia/Montenegro)